Programmed functional anatomy

PROGRAMMED FUNCTIONAL ANATOMY

GLYNN A. LEYSHON, Ph.D.

Associate Professor, Faculty of Physical Education,
University of Western Ontario; Associate
Professor, School of Medicine, University of
Western Ontario, London, Ontario, Canada

WITH 142 ILLUSTRATIONS

THE C. V. MOSBY COMPANY

SAINT LOUIS   1974

Library of Congress Cataloging in Publication Data

Leyshon, Glynn A   1929-
   Programmed functional anatomy.

   1. Anatomy, Human—Programmed instruction.
I. Title.   [DNLM: 1.   Anatomy—Programmed texts.
QS4 L685p   1974]
QM23.2.L5   611′.007′7   73-22279
ISBN 0-8016-2999-3

E/S/B   9   8   7   6   5   4   3   2   1

Preface

This programmed text in gross functional human anatomy was designed to give a basic overview of the subject to physical education students and to allied health students, such as physiotherapists and others with a similar interest. The emphasis given to the upper and lower limbs reflects this viewpoint.

The limbs are dealt with in their entirety rather than by system. In other words, the muscle, nerve, blood supply, skeleton, etc. are handled as integrated rather than as separate entities. This may lead to some difficulty initially, but it is hoped that it will ultimately leave the student with a more complete understanding of the parts involved and how these parts work together.

The portions of the body that are studied as systems were thought to lend themselves better to a separate rather than an integrated treatment; however, their functions in the integrated structure that constitutes a human were dealt with partially in the chapters covering the limbs. It is therefore wise to use the book in chapter sequence rather than at random. This is especially true of the early chapters in which basic nomenclature is covered.

As far as possible, the text and the illustrations have been correlated, and in some instances the illustrations have been used as questioning devices. The illustrations, like the rest of the book, are best used if supplemented with a good standard anatomy text such as *Introduction to Human Anatomy* by Carl C Francis.

Although prior learning in this field is not essential to the student using this text, a certain advantage is gained by the student who has taken a course in basic biology or a similar area of study.

Glynn A. Leyshon

Contents

1 Introduction to anatomical terms

The terms used in anatomy are universally accepted and agreed on so that anatomists can speak to one another in reference to the body without fear of confusion as to location or function of parts. The terms introduced in this chapter are used throughout the book and should be thoroughly mastered before proceeding with the "meat" of the text.

1 All descriptions in human anatomy are expressed in relation to the *anatomical position,* in which the body is erect, facing forward, with the palms of the hands facing forward. This anatomical position in which the palms face (forward/backward) is assumed even though the body is lying on its back (supine) or on its front (prone).

forward

2 The usual position for a body under study in a laboratory is on the back or in the (prone/supine) position.

supine

3 To make further references from the anatomical position, the body is divided into planes. A *median* plane is an imaginary line dividing the body into right and left halves. Thus a body in the prone position in which it is lying on its (front/back)

front

and divided down the middle on the _____ plane would have the (front/backs) of its hands uppermost.

median
backs

4 The vertical plane parallel to the median is called the *sagittal* plane. Thus a division of the body into two parts on a line running through one eye socket would be on a

_____ plane and running through the nose it would

sagittal

be on a _____ plane.

median

5 An examination of Fig. 1-1 will show two additional sections to the median and sagittal sections. They are the

_____ and _____ sections.

horizontal; frontal

6 The horizontal section runs at a right angle to the

_____ , _____ , and _____ planes.

median; sagittal; frontal
front and back

7 The frontal plane separates the body into (front and back/left and right) parts, whereas the median plane divides the body into (right and left/front and back) parts.

right and left

8 Although these terms are given in applying to the body as a

1

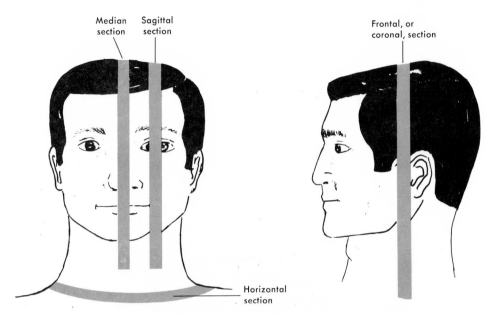

Median section   Sagittal section   Frontal, or coronal, section

Horizontal section

FIG. 1-1. Planes of dissection of the body.

whole, they apply to any part of the body as well. If a finger is sectioned into left and right halves, it is said to be a

median

_____ section of the finger.

9 A forearm with one third of its outer portion missing is said

sagittal

to be a _____ section of the forearm.

10 A forearm cut through at the elbow can be examined as a

horizontal

_____ section.

11 A finger cut through so that one third of the back of the

coronal

finger is removed is said to be a _____ section of the finger.

12 To refer to areas of parts of the body certain terms are used. The term *medial* means nearer to the median plane, whereas *lateral* means farther from it. In anatomical position the

lateral

thumb is (lateral/medial) to the little finger.

13 *Anterior* means nearer to the front of the body or limb, whereas *posterior* means nearer the back. Thus in anatomical

anterior
posterior

position the palm of the hand is (anterior/posterior), and the shoulder blades are (posterior/anterior).

14 *Superior* means nearer the top, and *inferior* means nearer the

inferior

lower end. In anatomical position the elbow is (superior/inferior) to the shoulder.

15 The terms *proximal* means closer to the attached end of a limb, whereas *distal* means farther away. Thus the thumbnail

distal

is at the (distal/proximal) end of the thumb.

2

internal; external

16 The terms *internal* and *external* mean nearer to and farther from the center of an organ or cavity. The heart is (internal/external), whereas the veins of the skin are (internal/external).

17 If something lies near or on the surface of the body, it is termed *superficial,* whereas something lying farther from the surface is *deep.* The bones of the arm are (superficial/deep) to the muscles attached to them.

deep

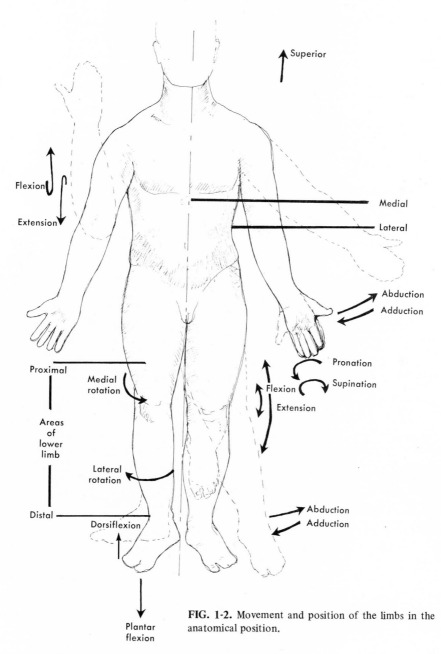

FIG. 1-2. Movement and position of the limbs in the anatomical position.

3

18 Movements of the limbs are also classified (Fig. 1-2). *Abduction* is movement of the limb in the coronal plane of the body away from the midline.

Adduction is the opposite, or a return to the midline. Lifting

abduction

the arm away from the side would be _____ and

adduction

returning it would be _____ .

19 *Flexion* and *extension* are movements in the sagittal plane. Flexion is the bending of a joint, and extension means straightening of a joint. Scratching one's head involves

flexing

bringing the hand up by (flexing/extending) the elbow joint. Returning the hand to the anatomical position involves

extending

_____ the elbow joint.

20 When dealing with the hip and neck joints, confusion may be avoided if one keeps in mind that flexing is bending and extension is straightening. Thus if one touches his toes, he is

flexing

_____ his trunk at the hips, and when he returns to

extending

the upright position, he is _____ the trunk at the hips.

21 If from the erect position the subject now bends backward, this is known as *hyperextension* of the back. If a gymnast does a backbend or a wrestler a bridge, he is demonstrating

hyperextension

_____ of the back.

22 The neck joint operates in a similar fashion. Putting the chin

flexing

on the chest constitutes _____ of the neck, and

extending

bridging is _____ of the neck.

23 *Rotation* occurs along the longitudinal axis of the rotating part. *Lateral* rotation occurs when the anterior aspect of the limb is turned away from the midline of the body. *Medial* rotation occurs when the anterior aspect is turned toward the midline of the body. If in the anatomical position the thumb is brought into contact with the thigh, the arm has been

medially

(medially/laterally) rotated. When the hand is returned to the

laterally

normal anatomical position, the arm must be (medially/ laterally) rotated.

24 *Pronation* and *supination* are types of rotation applied to the forearm and, rarely, to the foot as well. Pronation is medial rotation, and supination is lateral rotation of the forearm. When the palm of one's own hand is placed on a desk palm

medially

down, the forearm had to be (medially/laterally) rotated from

prone

the anatomical position to achieve this (supine/prone) position.

25 A special or different description of movements is applied to the foot. When a gymnast or diver points his toes, he is actually performing *plantar flexion* of the sole of his foot,

4

since the plantar surface is being flexed. Conversely, when he points his toes up to his face, he is performing *dorsiflexion*, since he is flexing the back, or dorsum, of the foot. In walking, a person (plantar flexes/dorsiflexes) his foot in the push-off phase.

plantar flexes

place kicking and leg
wrestling

26 The acts of (punting/diving/place kicking/leg wrestling) usually require dorsiflexion of the foot.

27 Describe the positions or movements depicted in Fig. 1-3.

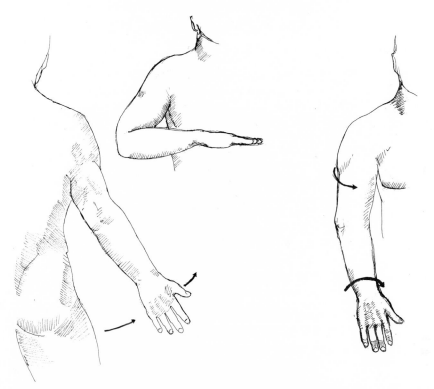

FIG. 1-3. Three movements of the limb. From right to left, abduction, flexion, rotation.

a. abducted

posterior

b. flexed

anterior

c. medially

a. The arm on the left is being _____ . The view is of the _____ surface of the body.

b. The arm in the center is in a _____ position. The view is of the _____ surface of the body.

c. The arm on the right is being _____ rotated.

28 The names of muscles are derived from various characteristics such as form, size, action, location, and number. The biceps brachii muscle of the arm is so named because the word "brachii" means arm and "bi" means two. Thus this muscle is named for its (action/location/size) and the (number/form/location) of its heads.

location; number

5

three	**29** A Latin prefix denoting a number is a part of naming muscles in many locations in the body. We have already noted that "bi" means two. Note also that "tri" means three and "quadri" means four. Thus the *triceps* of the arm has _____ heads, and the quadriceps of the thigh has
four	_____ heads.
action	**30** By the same token a muscle called the *pronator* is named for its (location/size/action).
action	**31** A muscle named the *adductor magnus* (magnus = big) is so called because of its size (magnus) and its _____ .
front location	**32** The term *tibialis anterior* describes a muscle of the leg (below the knee) that lies toward the (front/back) of the limb. It is therefore named for its (action/size/location).
trapezoid	**33** The terms *trapezius* and *deltoid* refer to muscles of a particular form. The first is shaped like a _____ (a geometrical figure), whereas the second is shaped like a triangle.
	34 With few exceptions muscles are attached to a bone at both ends. To distinguish one end from the other the more fixed end is termed the *origin*, and the more moveable is called the *insertion*. When a muscle contracts, it usually brings the
insertion	mobile end, the _____ , closer to the fixed end,
origin	the _____ .
	35 Muscles cannot push; they can only pull. If this precept is kept in mind, the tasks of most muscles can be deduced from
insertions	their origins and _____ .
	36 The elbow is flexed by the action of the biceps rather than by the muscles on the opposite surface, the triceps. When the
biceps	_____ flex the elbow, they pull the bones of the forearm toward the shoulder.
	37 This action makes the biceps the *prime mover* in flexion of
extension	the elbow. The return, or _____ , of the elbow is a function of the triceps. Since the muscles perform opposite actions, they are said to be *mutually antagonistic*.
	38 Most muscles in the body are set up in opposite acting pairs
mutually antagon- istic	such as the biceps and triceps, which are _____ _____ .
	39 For each prime mover, there is an *antagonist*. The antagonist
triceps	for the biceps is the _____ and vice versa.

6

$\mathcal{2}$ The shoulder

The skeleton of the upper limb consists of the bones of the fingers, hand, wrist, forearm, and upper arm (humerus), plus the clavicle (collarbone) and scapula. The latter two, not usually associated with the upper limb, are included because they provide a bony attachment, or connection, of the actual limb with the trunk.

1 The two bones not usually associated with the upper limb are the _____ and the _____ .

clavicle; scapula

2 These two bones are included because they provide a

_____

_____ .

bony attachment, or connection, of the limb to the trunk

3 The scapula (plural = scapulae) is roughly triangular in shape.

4 Examine Fig. 2-1. Is the apex of the triangle superior or inferior? _____

inferior

5 The base of the triangular scapula with its accompanying muscle layer forms what is commonly referred to as the _____ .

shoulder

6 The spine of the scapula is on the (anterior/posterior) surface.

posterior

7 The most lateral projection of the scapula is the _____

acromion

8 The scapula is concave (anteriorly/posteriorly).

anteriorly

9 The most forward, or anterior, process of the scapula is the _____ .

coracoid

10 The surface resting on the posterior surface of the ribs is the (concave/convex) surface.

concave

11 The glenoid fossa faces (medially/laterally) when the body is in normal anatomical position.

laterally

12 The upper portion of the scapula is very irregular to afford attachment to the many muscles involved in movement at the shoulder joint and to facilitate articulation with the *humerus* and *clavicle*. The irregularities

7

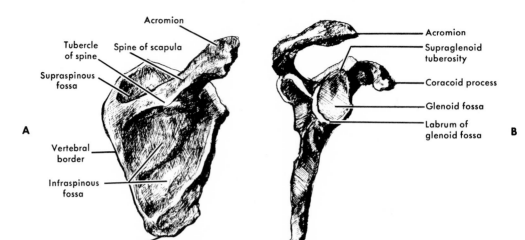

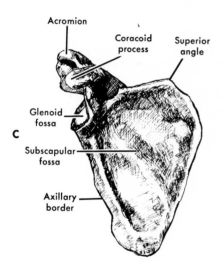

FIG. 2-1. The scapula. **A** shows the posterior view of the right scapula, **B** shows the lateral view, and **C** shows the anterior view.

muscles

of the scapula provide attachment for many _____ .

13 The scapula is connected to two bones, the

clavicle; humerus

_____ and the _____ .

14 Examine Fig. 2-2 and note the joining of the clavicle and scapula. The two parts of the scapula attached to the clavicle

acromion; coracoid

are the _____ and _____ processes.

15 Ligaments join bone to bone across a joint. In this case the ligaments involved in joining scapula and clavicle are the

acromioclavicular; cora-
    coclavicular

_____ and the _____ liga-
ments.

16 The head of the humerus fits into the glenoid fossa to form the *glenohumeral* joint, or shoulder joint. Thus the shoulder,

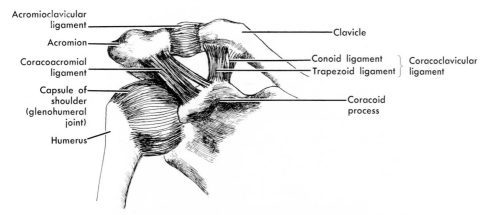

FIG. 2-2. Ligaments and capsule of the shoulder. Note the position of the ligaments that spring from the acromion and from the coracoid processes. The capsule has some slight thickenings that act as weak ligaments of the shoulder joint.

glenohumeral joint	or the _____ , is composed of two bones,
humerus; scapula	the _____ and the _____ . The bones are joined by a thick *capsule*.
	17 The position of the capsule of the glenohumeral joint
ligament	suggests that it acts as a _____ to join the
scapula	humerus to the _____ .
	18 The capsule has thickenings, or folds, that act as weak ligaments. The joint is supported by ligaments farther from the capsule. The coracoacromial ligament joining the
coracoid; acromion	_____ process to the _____ process provides an arch, or roof, over the glenohumeral joint and the tendon of the *supraspinatus* muscle.
coracoid	**19** The coracoclavicular ligament joining the _____
clavicle	process and the _____ reinforces the acromioclavi-
acromion	cular ligament joining the _____ process and the
clavicle	_____ .
	20 In a fall on the outstretched arm, the acromion is sometimes driven under the clavicle (an acromioclavicular separation). The trapezoid ligament, a portion of the larger
coracoclavicular	_____ ligament, resists this movement of the acromion.
arch	**21** The coracoacromial ligament serves as an _____ over
humerus	the head of the _____ and over the tendon of the
supraspinatus	_____ muscle.

9

22 The coracoclavicular ligament serves to protect against an

acromioclavicular

_____ separation in a fall.

23 There are then three ligaments at the shoulder: The two-part

coracoclavicular; acro-
mioclavicular; cora-
coacromial

_____ , the _____ , and the

_____ ligaments. In addition, the glenohumeral

capsule

joint is supplied with a _____ .

24 You can palpate your own clavicle readily. Start at the
medial end where it is attached to the *sternum* (breastbone),
and note that there is a considerable gap between the

medial

(medial/lateral) ends of your right and left clavicles.

25 Judging from the way names of other joints were derived
from the bones they connect, the name of the joint between

sternoclavicular

sternum and clavicle is the _____ joint.

26 Supporting the sternoclavicular joint, as with other joints, are

ligaments

several _____ .

sternoclavicular

27 One of the ligaments is named the _____
ligament, since it joins the sternum and clavicle.

28 There is a sternoclavicular ligament both at the front, or

anterior

_____ , surface and on the back, or

posterior

_____ , surface of each bone. In addition, there
is an *interclavicular* ligament and also a *costoclavicular* ligament running from the first rib to the clavicle (Fig. 2-3).

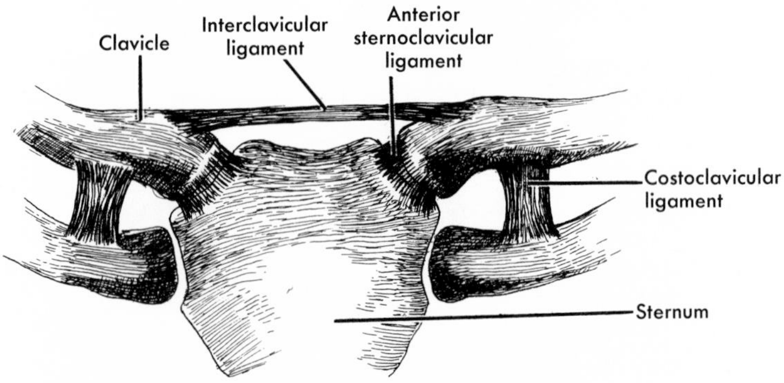

FIG. 2-3. Ligaments of the sternoclavicular joint.

29 The clavicle is attached at its medial end to the

sternum; two

_____ . There are (three/two) ligaments effecting

costoclavicular

this union, plus another called the _____
ligament, which is attached to the first rib.

30 If you run your fingers along your right clavicle from medial

lateral; acromion

to lateral, you will note a slight crevice between the (lateral/medial) end of the clavicle and the (acromion/coracoid) process, which forms the hard "point of the shoulder."

acromioclavicular

31 The crevice marks the site of the _____ joint.

32 Further run your fingers from the acromion, over the shoulder slightly, and then medially until you strike a bony crest, or ridge. It is part of the scapula. The bony crest

spine

palpable in this area is the _____ of the scapula.

vertebral

33 Medially, the vertical hard edge palpable is the _____ border of the scapula.

34 The movement of the scapula is extensive and because the scapula can be moved to so many positions, the

glenoid

_____ fossa can face in many different positions; this, in turn, permits the upper limb a great range of motion at the shoulder. Note the medial-lateral swing of the scapula in Fig. 2-4.

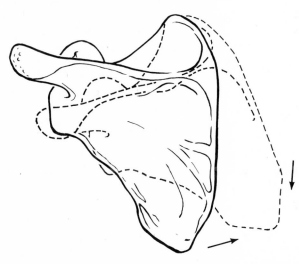

FIG. 2-4. Rotation of the scapula. This rotation turns the glenoid fossa up, forward, or down and gives the arm a great range of motion.

35 The great range of motion of the shoulder joint is possible

scapula

because the _____ rotates.

36 If you cross your outstretched arms in front of you and pull

clavicles

them medially, two bones, the _____ , limit the medial movement of the humeri because they act as struts to keep the humeri in position.

37 There are five movements of the scapula as it slides over the thoracic wall. Two of these are *elevation,* in which the shoulder is raised, and *depression,* in which the shoulder is

11

returned from the elevated position. When one shrugs his

elevates

shoulders, he _____ them. The opposite action is

depression

called _____ , and it lowers them.

38 Two additional movements are protraction and retraction. *Protraction* is the moving of the scapula forward and laterally, whereas *retraction* of the shoulders draws the scapula back in military fashion. Standing at rigid attention

retracted

calls for the scapula to be _____ . Assuming the position of a quarterback under a center calls for the scapula

protracted

to be _____ .

39 Finally, the fifth movement of the scapula is rotation. The scapula rotates on an imaginary pivot point near the lateral end of the spine of the scapula so that the inferior angle can swing laterally or medially, thus turning the glenoid fossa upward or downward. When a gymnast reaches up to grasp

laterally

the still rings, his scapulae rotate (medially/laterally), and his

upward

glenoid fossae face (upward/downward).

40 In summary, the movements of the scapula are

elevation; depression

_____ , _____ ,

protraction; retraction

_____ , _____ ,

rotation

and _____ .

41 In movements such as raising the upper limb overhead, there

three

are (one, two, three) bones involved at the shoulder.

clavicle

42 The bones involved at the shoulder are the _____ ,

scapula; humerus

_____ , and _____ .

43 Since the scapula moves, and since the clavicle is attached to it firmly, it follows that the clavicle is moveable at its

medial

(lateral, medial) end.

44 Both the glenohumeral joint and the sternoclavicular joint are synovial joints. There are four characteristics of a synovial joint:

a.   The joint is encased in a capsule.

b.   There is often a fibrocartilage disc between the two articulating surfaces.

c.   The joint is bathed in a fluid called *synovium,* which provides lubrication.

d.   The joint is freely moveable. The acromioclavicular

is not

joint (is, is not) a synovial joint.

45 The glenohumeral (shoulder) joint (Fig. 2-2) is identifiable

capsule

as a synovial joint by the presence of (fibrocartilage/capsule) extending from the shaft of the humerus to the scapula.

46 Since the medial end of the clavicle moves when the scapula is moved, the sternoclavicular joint must be a

synovial

_____ joint.

47 The synovium bathing the glenohumeral and sternoclavicular

lubricant

joints acts as a _____ .

MUSCLES OF THE ROTATOR CUFF

48 Among the muscles of the shoulder girdle that work to move
the upper limb are four known collectively as the rotator
cuff muscles. The first of these muscles of the

rotator

_____ cuff group is the *supraspinatus.*

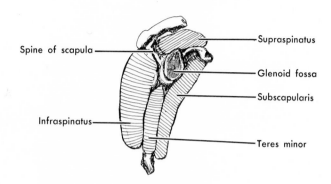

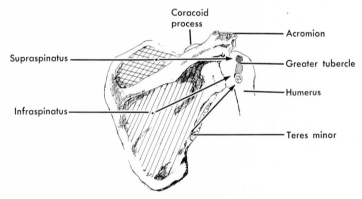

FIG. 2-5. Site and attachment of the rotator cuff muscles.

49 The supraspinatus muscle originates in the supraspinous
fossa (Fig. 2-5) and inserts on the greater tubercle of the

superior

humerus. The supraspinatus is (superior/inferior) to the spine
of the scapula.

50 The *infraspinatus* originates on the scapula in the infra-

posterior

spinous fossa on the (anterior/posterior) surface, and it
inserts on the same site on the humerus as does the

greater tubercle

supraspinatus. The point of insertion is the _____

_____ , although slightly lower than that of the
supraspinatus.

51 The third muscle of the group is the *teres minor.* It arises on

13

lateral

supraspinatus; infra-
    spinatus

under

the axillary border of the scapula and is (lateral/medial) to the infraspinatus. It inserts onto the greater tubercle in a position lower than the other two muscles attached there, the _____ and the _____ .

52 The last muscle of this group is the *subscapularis.* As its name implies, the muscle originates (under/atop/beside) the scapula and inserts on the lesser tubercle of the humerus (Fig. 3-1).

53 The four named muscles are often recalled by the mnemonic SITS, derived from the first letter of each. The "I", for

infraspinatus

example, stands for the _____ muscle.

54 The two muscles of the group separated by the spine of the

supraspinatus

infraspinatus; supra-
    spinatus

scapula are the _____ and the _____ , the _____ being the superior.

subscapularis

55 The _____ arises from the anterior surface of the scapula.

56 The muscles all insert into the greater tubercle of the

subscapularis

humerus with the exception of the _____ , which

lesser

inserts into the _____ tubercle.

57 The SITS muscles as a group not only act on the humerus but also reinforce the rather thin, weak capsule of the shoulder joint and help supply the support at the joint

ligaments

normally supplied in other joints by the _____ .

58 The most important job of the muscles of the rotator cuff is to hold or stabilize the head of the humerus in place when another muscle, the *deltoid,* raises or abducts the upper limb. Their action has been likened to that of a man holding one end of a ladder down while a companion raises the opposite end and walks toward him gradually raising the ladder. If one end were not held, the ladder could not be raised to a vertical position. The rotator cuff holds the head of the humerus just as one holds the end of a ladder. (See Fig. 2-6.)

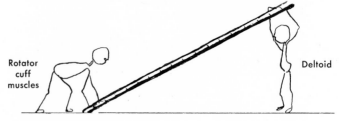

Rotator
cuff
muscles

Deltoid

FIG. 2-6. The rotator cuff muscles act to stabilize the head of the humerus during abduction of the arm.

59 The muscles of the rotator cuff not only reinforce the

stabilize capsule but also serve to _____ the head of the

abduction humerus in the movement of _____.

60 The main or prime mover in abduction of arm is not the

deltoid rotator cuff muscles but the _____ muscle.

61 Examine Fig. 2-5. From the origins and insertions of the rotator cuff muscles one could deduce that they function as follows:

abduction *supraspinatus*—aids in (abduction/adduction)
laterally *infraspinatus*—rotates the arm (laterally/medially)
laterally *teres minor*—rotates the arm (laterally/medially)
medially *subscapularis*—rotates the arm (laterally/medially)

62 If you were standing in the anatomical position and you turned your hand so that your thumb touched your thigh,

medial the muscle producing this (medial/lateral) rotation would be

subscapularis the _____ .

63 When you return your thumb to its original position, the

infraspinatus; teres _____ and _____ act to pro-
minor
lateral duce this _____ rotation.

64 The inferior part of the shoulder joint area has less muscle, ligament, and capsule than the rest. The joint is more susceptible to dislocation, or *subluxation,* inferiorly than in any other direction. This is especially true when the arm is abducted and laterally rotated as in the overhand striking action. The action of spiking in volleyball is dangerous for a person with a chronic shoulder problem because the arm is

abducted; laterally (abducted/adducted) and (medially/laterally) rotated.

65 The act of spiking a volleyball might cause a _____

dislocation (sub- _____ of a chronically
luxation)
troubled shoulder.

66 The most common direction of dislocation of the shoulder is

inferiorly (medially/laterally/inferiorly) because the joint is weakest in this aspect.

67 The innervation of the cuff muscles is an follows:
supraspinatus—suprascapular nerve
infraspinatus—suprascapular nerve
teres minor—axillary nerve
subscapularis—subscapular nerve

suprascapular The _____ nerve innervates two muscles of the
supraspinatus; infra-
spinatus rotator cuff, the _____ and _____ .
medially 68 When called on to (laterally/medially) rotate the arm, the

subscapular subscapularis is activated by the _____ nerve.

15

69 The axillary nerve stimulates the _____
to (medially/laterally) rotate the arm.

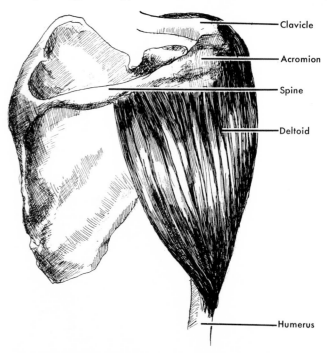

Clavicle

Acromion

Spine

Deltoid

Humerus

FIG. 2-7. The deltoid, a bulky triangular muscle shaped like an inverted Greek letter "D"—hence its name.

DELTOID MUSCLE

70 The deltoid muscle (Fig. 2-7) is the chief (abductor/adductor) of the shoulder. It originates on the clavicle plus acromion and spine of the scapula.

71 From its origin on the spine of the scapula and on the

_____ and the _____ , the deltoid inserts on the deltoid tuberosity of the humerus.

72 The point of the insertion on the humerus, called the

_____ , is on the (lateral/medial) surface about halfway down the shaft of the humerus.

73 Palpate your own deltoid. It gives your shoulder its rounded contour and can be most easily felt when your arm is abducted.

74 The deltoid is innervated by the same nerve supplying the

teres minor, the _____ .

75 The three parts of the deltoid, the clavicular, acromial, and

spinous, are named for their respective sites of (origin/insertion) and perform different actions.

16

middle	**76** The acromial is the (anterior/posterior/middle) portion of the deltoid and is responsible for abduction.
	77 Whereas the acromial portion is responsible for
abduction; clavicular	_____ , the anterior, or (clavicular/spinous), portion gives the action of flexion of the shoulder and also medially rotates the arm. Flexing the shoulder raises the arm in front such as in throwing an upper cut or serving in badminton.
spinous	**78** The remaining part of the deltoid, the _____ section, enables one to extend his shoulder and laterally rotate the arm. Extending the shoulder raises the arm to the rear such as when pumping the elbows while sprinting.
	79 Thus the deltoid by means of its three portions is capable of
five	(one, two, three, four, five) movements.
flexion	**80** The clavicular portion is responsible for _____
medial	and _____ rotation.
extension	**81** The spinous portion is responsible for _____
lateral	and _____ rotation.
acromial	**82** The middle, or _____ , portion is responsible
abduction	for _____ .
	83 Slowly lowering a weight held in the outstretched arm will also bring the deltoid into play because the action opposite
abduction	to the lowering, or adduction, is _____ .
three	**84** The deltoid has (four/two/three) major portions; it is
axillary	innervated by the _____ nerve and performs the following five actions:
flexion; extension abduction; medial rotation; lateral rotation	_____ , _____ , _____ , _____ , and _____ .
	85 The Greek letter "delta" is shaped like a triangle. The deltoid muscle, although inverted, is so named because of its resemblance to the Greek letter. The apex of the triangular
humerus	deltoid is attached to the _____ .

TERES MAJOR MUSCLE

	86 The name *teres minor* would suggest that there is a related
major	muscle, a teres _____ .
	87 Examine Fig. 2-8. The teres major is slightly larger than the
inferior	minor and is (inferior/superior) to it in its origin on the
scapula	_____ .

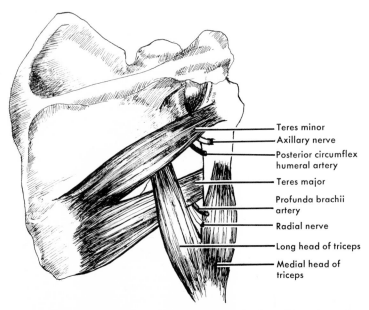

Teres minor
Axillary nerve
Posterior circumflex humeral artery
Teres major
Profunda brachii artery
Radial nerve
Long head of triceps
Medial head of triceps

FIG. 2-8. Relationship of the teres major and minor. Note the major nerves and arteries of the posterior aspect of the arm.

88 Although the teres major is inserted on the same bone, the

humerus   _____ , as the teres minor, it inserts on the anterior surface of the shaft at the lesser tubercle.

89 From its position one could deduce that the teres major acts

medial   to cause (medial/lateral) rotation of the humerus. It also

adducts   (abducts/adducts) the humerus, as well as extends the shoulder.

90 The nerve supply to the teres major is via the *subscapular*

axillary nerve   nerve, whereas that of the teres minor is the _____

_____ .

91 In performing a chin-up, the teres major would work to help

extend   (flex/extend) the shoulder, and it would be stimulated by

subscapular   the _____ nerve.

TRAPEZIUS MUSCLE

92 Like the deltoid, another large muscle attached to the

superficial   scapula is (superficial/deep). It is called the *trapezius* and is considered a muscle of the back.

three   **93** Also like the deltoid, the trapezius has (one/two/three) rather distinct portions that perform different functions. Examine Fig. 2-9 and note the attachments of the trapezius.

18

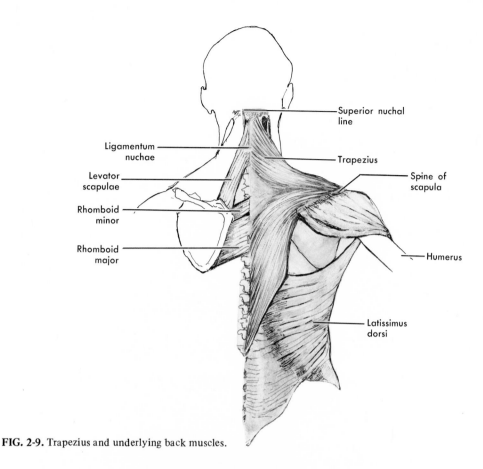

FIG. 2-9. Trapezius and underlying back muscles.

scapular

clavicular

spine

clavicle

ligamentum nuchae

spine

thoracic

three

94 The deltoid originates on the _____ and _____ bones, whereas the trapezoid *inserts* on the same two bones.

95 The trapezius inserts on the lateral third of the clavicle and on the acromion. From Fig. 2-9 it is evident that the trapezius shares the (spine/coracoid) of the scapula with the deltoid.

96 The trapezius originates on the *superior nuchal line* (Fig. 2-9) of the skull, and this portion goes to the most anterior insertions on the _____.

97 The next lower portion of the origin of the trapezius is indicated as the _____ on Fig. 2-9. This part of the trapezius inserts on the acromion and the _____ of the scapula.

98 The lowest fibers of the trapezius originate on the _____ vertebrae (all twelve of them) and insert onto the spine of the scapula.

99 The trapezius actually has (one/three/two) areas of origin,

nuchal line

twelfth

stretching from the _____ on the skull down to the (eighth/tenth/twelfth) thoracic vertebra.

100 The trapezius is superficial and may be palpated readily (Fig. 2-9). Palpate your own while moving your scapula, that is, shrugging your shoulders. What action does the upper

elevation, or hunching

portion provide? _____

101 The middle fibers from their position most likely cause the

retract

scapula to (retract/protract).

102 The inferior fibers also serve, as the middle fibers do,

retract

to _____ the scapula.

103 The superior fibers serve to rotate the scapula upward during abduction of the arm. In reaching for a rebound, a basketball player rotates his scapula upward with the aid of

trapezius

his _____.

104 The trapezius is innervated by the *accessory* nerve. Thus in

superior

shrugging one's shoulders the (superior/inferior) fibers of the

trapezius; accessory

_____ are stimulated by the _____ nerve.

RHOMBOID AND LEVATOR SCAPULAE MUSCLES

105 If the trapezius is parted and reflected on a specimen, visible

deep

(superficial/deep/medial) to it are three more muscles attached to the scapula. These are *rhomboid* major and minor and *levator scapulae* muscles (Fig. 2-9).

106 The three muscles are all inserted onto the (medial/lateral/

medial

inferior) border of the scapula.

107 The levator scapulae, judging from its name, must (depress/

elevate

elevate) the scapula much as the upper portion of the trapezius does.

108 The rhomboids act with the middle portion on the trapezius

retract

to _____ the scapula.

109 Like the teres minor of the pair of teres muscles already

superior

discussed, the rhomboid minor is (superior/inferior/lateral) to the rhomboid major.

110 The rhomboid major, being situated (deep/inferior/lateral)

inferior

to the other rhomboid, originates on the spines of the second to fifth thoracic vertebrae.

111 The rhomboid minor originates on the spines of the seventh (and last) cervical (= neck) and first thoracic vertabrae. Its

less

origin is (more/less) extensive than the major.

112 The rhomboids are innervated by the *dorsal scapular* nerve.

retracting

Thus in the action of (protracting/retracting) the scapula, the

rhomboid major

trapezius acts with the _____

rhomboid minor

muscle and _____ muscle, the

20

dorsal scapular

latter two being stimulated by the _____ nerve.

113 The levator scapulae arises from the transverse processes of the first four cervical vertebrae. Thus it is close in origin to the (lower/upper/middle) fibers of the trapezius. It inserts above the rhomboid minor on the (medial/lateral) border of the scapula.

upper
medial

114 The levator scapulae is innervated by the third and fourth cervical nerves. When shrugging your shoulders, two muscles,

levator scapulae; tra-
pezius

the _____ and _____ , operate. They are

third; fourth

stimulated by the _____ and _____

accessory

cervical nerves and the _____ nerve, respectively.

SERRATUS ANTERIOR MUSCLE

medial

115 The *serratus anterior,* so named because of its saw-toothed, or serrated, appearance, arises from the upper eight ribs and attaches to the (medial/lateral/superior) border of the scapula (Fig. 2-10).

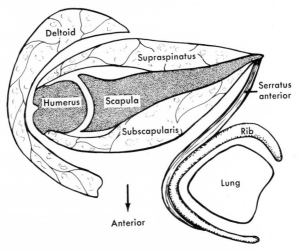

FIG. 2-10. Horizontal section of the right shoulder.

116 Its insertion extends from the superior angle to the inferior angle of the scapula, and it lies (anterior/posterior) to the subscapularis. About half the muscle is inserted at the inferior angle of the scapula.

anterior

117 When the serratus anterior contracts, it draws the inferior angle (medially/laterally/superiorly).

laterally

118 By drawing the inferior angle laterally, the serratus anterior aids in (abduction/adduction/rotation) of the arm.

abduction

21

119 Sometimes called the "fencer's muscle" because it pulls the scapula forward in the motion of thrusting, the serratus anterior is innervated by the *long thoracic* nerve. In fencing, a competitor is likely to use his serratus anterior when he

thrusts

_____ . In this case the muscle is stimulated by

long thoracic

the _____ nerve.

BLOOD SUPPLY TO THE SHOULDER

120 The blood supply to the scapular area is carried by branches from the subclavian artery.

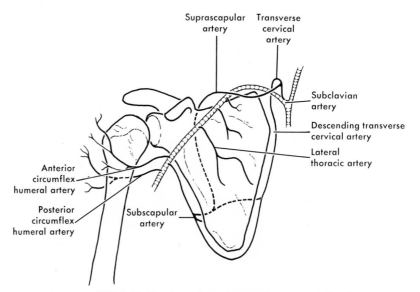

FIG. 2-11. Blood supply to the scapular area, anterior view.

121 Examine Fig. 2-11. Which artery would most likely supply

suprascapular

supraspinatus and the infraspinatus? _____

122 With which artery does the suprascapular artery anastomose?

subscapular

_____

123 The rhomboids and levator scapulae would receive their

transverse cervical

blood supply from the descending _____

serratus anterior

_____ artery, as would the _____

_____ muscle, which also attaches to the medial border of the scapula.

lateral thoracic

124 The _____ artery lies on the anterior surface of

subscapularis

the scapula and supplies the _____ muscle, which arises on the same surface of the scapula.

22

125 The deltoid is probably supplied by the two circumflex

anterior circumflex

posterior circumflex

arteries, the _____ humeral

artery and the _____ humeral artery.

NERVE SUPPLY TO THE SHOULDER

126 The shoulder joint itself is innervated according to *Hilton's law,* which states that a joint is innervated by the same nerves that stimulate the muscles crossing the joint. In this case the following nerves would stimulate the shoulder joint:

a. suprascapular

a. _____ nerve because it innervates the supraspinatus and infraspinatus

b. subscapular

b. _____ nerve because it innervates the subscapularis and teres major

c. axillary

c. _____ nerve because it innervates the deltoid and teres minor

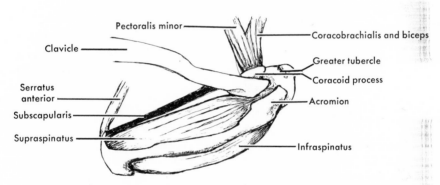

FIG. 2-12. Top view of coracoid and its muscle attachments (right side).

MUSCLES ARISING ON THE CORACOID PROCESS

127 The muscles attached to the coracoid process (crow's beak) are the *pectoralis minor,* and *coracobrachialis,* and the short head of the *biceps.* The coracoid process (Fig. 2-12) is

inferior

three

(superior/inferior) to the clavicle.

128 There are (one/two/three) muscles attached to the coracoid process.

129 The coracoid process is named for its resemblance to a

crow's beak

_____ .

130 The pectoralis minor is a muscle of the chest, whereas the

coracobrachialis

short head of the biceps

remaining two muscles, the _____ and the

_____ ,

insert on the humerus (brachium = arm).

23

131 The nerve supply to the coracobrachialis (running from the

coracoid process

_____ to the humerus) and the
short head of the biceps is the *musculocutaneous* nerve.
Thus when the arm is flexed at the shoulder, the anterior

coracobrachialis

deltoid is assisted by two muscles, the _____

short head of the biceps

and the _____ ,

musculocutaneous

which are stimulated by the _____ nerve.

132 The term *short head of the biceps* implies that there is a

long

_____ head of the biceps.

MUSCLES ARISING FROM THE RIM
OF THE GLENOID FOSSA

133 The long head of the biceps arises from the *supraglenoid
tubercle* of the scapula and crosses the shoulder joint. Since

flexing

it lies anteriorly, it must assist the deltoid in (flexing/
extending) the shoulder (Fig. 3-3).

134 The muscle that supplies a function opposite to the long
head of the biceps is the long head of the *triceps*. It
originates on the *infraglenoid tubercle* and lies on the
posterior surface of the humerus. It helps the deltoid

extend

(flex/extend) the shoulder (Fig. 3-4).

REVIEW OF THE MAJOR MUSCLES OF THE SHOULDER

135 The most easily palpated shoulder muscles are the

deltoid; trapezius

_____ and the _____ .

136 The most anteriorly placed muscle attached to the scapula is

serratus anterior

the _____ .

137 The muscles attached to the axillary, or lateral, border are

teres minor; teres major

the _____ and _____ .

138 The muscles attached to the medial, or vertebral, border
from superior to inferior aspect are the:

levator scapulae

_____

rhomboid minor

_____

rhomboid major

_____

serratus anterior

_____

BRACHIAL PLEXUS

139 The area of the *axilla,* or armpit, is important because of the
presence of a bundle of nerves known as the *brachial plexus,*
branches of which innervate nearly all the muscles thus far
studied and all those of the upper limb. When one abducts

his arm, for example, the muscles causing the movement are stimulated by the axillary and other nerves that arise from

brachial plexus

the _____. Examine Fig. 2-13 and note the major terminations (boldface type).

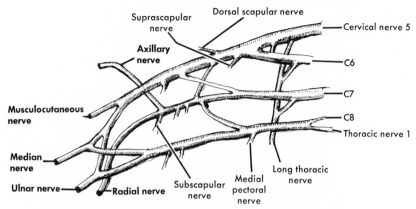

FIG. 2-13. Diagrammatic representation of the brachial plexus.

five

140 There are (one/three/five) major terminating branches of the brachial plexus.

141 There are six nerves shown in Fig. 2-13 that have been mentioned thus far in connection with muscles of the shoulder:

a. supraspinatus

  infraspinatus

a. suprascapular, which innervates the _____ and _____ muscles

b. rhomboids

b. dorsal scapular, which innervates the _____

c. deltoid

  teres minor

c. axillary, which innervates the _____ and _____

d. serratus anterior

d. long thoracic, which innervates the _____

e. short head of the bi-
  ceps; coracobrachialis

e. musculocutaneous, which innervates the _____ _____ and _____

f. subscapularis

  teres major

f. subscapular, which innervates the _____ and _____ muscles

(NOTE: a nerve may innervate more than one muscle.)

posterior

142 Examine Fig. 2-14 and note that the branches lie (anterior/posterior/lateral) to the pectoralis minor.

25

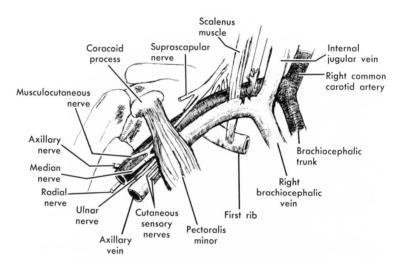

FIG. 2-14. Brachial plexus in relation to other structures.

3 The arm

1 The bone correctly referred to as the bone of the arm is the

humerus   _____.

2 It articulates at its proximal end with the glenoid fossa of the

scapula   _____.

3 At its distal end the humerus articulates with two bones, the radius and the ulna. Thus the humerus articulates with a total

three   of (three/two/one) bones, the more distal articulation being

radius; ulna   with the _____ and _____ .

4 On Fig. 3-1, note the point of the insertion of the deltoid

midway   muscle on the deltoid tuberosity. It is approximately (midway/one-fourth way/one-third way) down the humerus on

lateral   the (anterior/medial/lateral) surface.

anterior   **5** The bicipital groove on the (anterior/posterior/lateral) surface contains the tendon of the long head of the biceps.

distal   **6** There are two fossae on the (distal/proximal) end of the

posterior   humerus, the *olecranon* fossa on the (anterior/posterior)

coronoid   surface and the _____ fossa on the opposite surface.

7 The *capitulum* articulates with the head of the radius, and the *trochlea* articulates with the remaining bone of the forearm,

ulna   the _____.

8 Palpate your own humerus at its distal end and locate the medial and lateral epicondyle (= above the knuckle). Note that the lateral epicondyle is superior to and lateral to the

capitulum   (capitulum/trochlea).

9 The muscles that flex and extend the elbow originate on the shoulder girdle and on the humerus. Since they act on the

radius   elbow joint, they must insert on the _____

ulna   and _____.

10 The muscles are divided into two groups. Generally speaking, the group on the anterior surface of the humerus causes

flexion   (flexion/extension) of the elbow, and the group on the

extension   posterior surface causes (flexion/extension) of the elbow. Examine Fig. 3-2 and note the two groups.

27

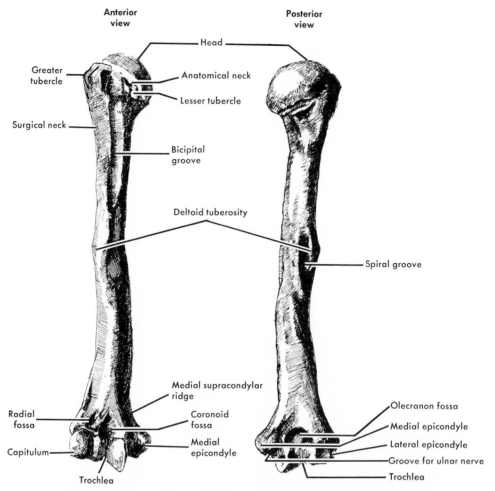

FIG. 3-1. Anterior and posterior views of a right humerus. Note that the bicipital groove is also known as the intertubercular groove.

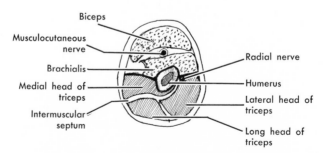

FIG. 3-2. Compartments of the arm. Tough fascia separates the anterior from the posterior compartment. The coracobrachialis is not shown because the section was made below the point on the humerus where the short coracobrachialis inserts.

11 The groups, or compartments, are separated by intermuscular septa composed of tough fascia.

ANTERIOR COMPARTMENT

12 The following are the muscles of the flexor, or anterior, compartment. Their origins have been discussed previously, or they can be deduced from their names:

a. The long head of the biceps, originating on the

a. supraglenoid

_____ tubercle of the scapula

b. The short head of the biceps, originating on the

b. coracoid

_____ process of the scapula

c. coracoid

c. The coracobrachialis, originating on the _____ process of the scapula

d. The *brachialis*, originating on the humerus

13 The biceps brachii (long and short heads) is the muscle most commonly referred to when speaking of the flexors of the elbow. It is a flexor of the elbow and has a common tendon

insertion

of (insertion/origin) on the radial tuberosity and also into the deep fascia of the forearm.

anterior

14 The biceps brachii, the most superficial muscle of the (anterior/posterior) compartment, is inserted directly into the tuber-

radius

osity of the _____ and, by way of the fascia of the forearm, into the remaining bone of the forearm, the

ulna

_____ .

15 The coracobrachialis, as its names implies, arises from the

coracoid; humerus

_____ and inserts on the _____, about the middle third of the medial border.

humerus

16 The brachialis arises on the _____ along the distal two thirds. It origin embraces the insertion of a large

deltoid

triangular muscle of the shoulder, the _____ .

17 The brachialis inserts into the capsule of the elbow joint and

deep

the tuberosity of the ulna. It is (superficial/deep) to the biceps.

18 The innervation of the muscles of the anterior compartment is simple to remember. They are all supplied from the nerve

musculocutaneous

that supplies the biceps, the _____ nerve.

19 For a muscle to create movement at a joint, it must cross the joint. Thus when one looks for the muscles that flex the

biceps brachii

elbow joint, he must include the _____ because the tendon crosses the joint.

20 When a weight trainer practices curls (elbow flexion), he calls

two

on at least (five/three/four/two) muscles to perform the

29

exercise (count biceps brachii as one muscle). The muscles are

biceps brachii; brachialis   the _____ and _____
(Fig. 3-3).

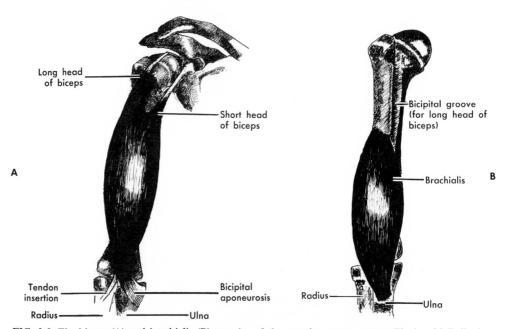

FIG. 3-3. The biceps (A) and brachialis (B) muscles of the anterior compartment. The brachialis lies beneath the biceps.

21 Note on Fig. 3-3 the length of the brachialis compared to the

ulna

biceps. The brachialis, which inserts on the _____ , has been called the workhorse of the flexors because it provides most of the force when flexion is performed against heavy resistance.

one
one

Unlike the biceps brachii, the brachialis is a (one-/two-/three-) joint muscle, that is, it crosses (one/two/three) joints.

flexing

22 The biceps, in addition to its job of (flexing/extending) the elbow, performs another action on the forearm because of the unique fascial insertion known as the *bicipital aponeurosis* on the ulna. If the elbow is in semiflexed position, and the palm is down, the ulna can be drawn atop the radius to complete

supination

the action of (supination/pronation). Such an action is observed in rotating a doorknob or using a screwdriver to drive a screw.

POSTERIOR COMPARTMENT

three

23 The extensors of the elbow are the triceps brachii. From its name, you could surmise it has (one/two/three) heads of origin.

30

24 The *long head* of the triceps arises from the infraglenoid

long head of the biceps   tuberosity. In a sense it is opposite to the _____

_____ of the flexor group, which

supraglenoid   arises from the _____ tuberosity. Note on
Fig. 3-4 the common tendon of insertion for all three heads
of the triceps.

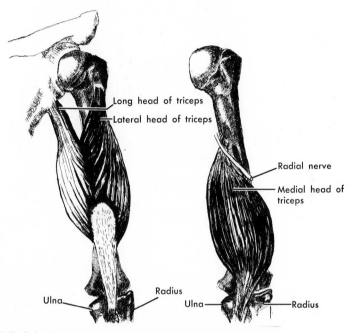

FIG. 3-4. The triceps. The medial head lies deep to the long and lateral
heads.

25 The *lateral head* of the triceps arises from the upper portion
of the posterior humerus. It inserts in common with the

long   _____ head and the medial (deep) head on the
olecranon of the ulna (Fig. 4-1) and into the fascia of the
posterior forearm.

deepest   **26** The *medial head* is the (deepest/most superficial) of the three
heads; it originates on the posterior humerus and inserts on
the olecranon in common with the other heads.

27 The triceps are innervated by branches of the *radial* nerve.
Thus when you attempt to do a push-up or throw a ball, the

radial   _____ nerve stimulates the

triceps; extension   _____ , and (flexion/extension) of the elbow
results.

28 The medial head acts in routine extension of the elbow, whereas the long and lateral heads are called to work when the resistance to extension is greater. The only head of the

long triceps that is a two-joint muscle is the _____ head.

two **29** In summary, one could say that there are (one/two/three) compartments in the upper arm, containing a total of

six _____ muscles.

posterior **30** The muscles of the (anterior/posterior/medial) compartment are innervated by the radial nerve and its branches, whereas

anterior those of the (medial/anterior/posterior) compartment are

musculocutaneous innervated by the _____ nerve.

ADDUCTORS OF THE ARM

31 A strong adductor of the arm working in opposition to the

deltoid strong abductor of the arm, the _____ muscle, is the *pectoralis major,* originating on the upper six ribs and the clavicle (Fig. 10-4).

upper six ribs **32** From its origin on the _____and

clavicle _____ the pectoralis major inserts onto the greater tubercle of the humerus and upper anterior shaft.

greater tubercle **33** Since the pectoralis major inserts onto the _____

_____ from its origins on the clavicle and ribs, its tendon twists.

34 The upper, or clavicular, portion of the pectoralis major not only adducts the arm but also elevates it, and the lower

upper six ribs portion arising from the _____ depresses the arm and shoulder. Pressing down on a tabletop employs the sternal portion, and lifting up on the tabletop employs the clavicular portion.

medial **35** Also, the pectoralis major is important in (medial/lateral) rotation of the arm and is innervated by the *lateral* and *medial pectoral* nerves.

all three **36** The pectoralis major is important in the action of (throwing/

lateral and medial pushing/shoveling) under stimulation from the
    pectoral

_____ nerves.

37 A smaller muscle, the *pectoralis minor,* lies deep to the

behind pectoralis major. In its position (behind/in front of) the pectoralis major, the pectoralis minor arises from the second to sixth ribs.

second to sixth ribs **38** From its origins on the _____the pectoralis minor inserts onto the coracoid process.

39 The insertion of the pectoralis minor onto the

coracoid process _____ enables it to depress the shoulder.

lateral and medial
pectoral

40 The pectoralis minor is supplied by the same nerves supplying the pectoralis major, the _____

_____ nerves.

41 The *latissimus dorsi* (Fig. 10-4) is a strong adductor of the arm originating on the lower back. It works in concert with

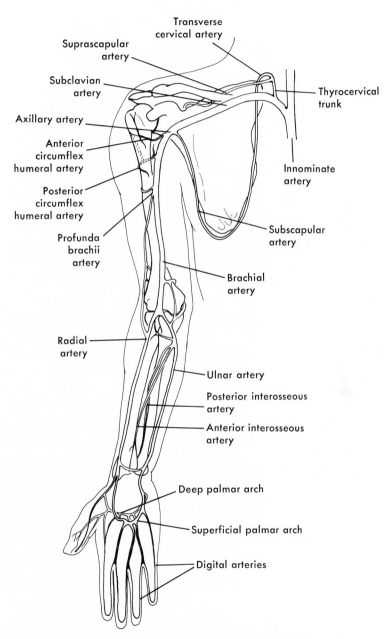

FIG. 3-5. Summary of the circulation to the upper limb. Note the many anastomoses around the shoulder, elbow, and wrist joints.

pectoralis major	the adductor from the anterior body surface, the _____ _____.
lower back	**42** The origin of the latissimus dorsi on the _____ encompasses the spine and some fascia before the fibers run up to the medial lip of the bicipital groove.
anterior	**43** From its insertion in the medial lip of the bicipital groove on the (anterior/posterior) surface of the humerus, the latissimus dorsi is able not only to adduct the arm but also to extend it.
medial lip of the bicipital groove	**44** The insertion on the _____ is small compared to the vast origin of the latissimus dorsi and contributes to making the muscle triangular. It is innervated by the *thoracodorsal* nerve in its actions of
adducting; extending	_____ and _____ the humerus.

BLOOD SUPPLY TO THE UPPER LIMB

	45 Examine Fig. 3-5 and note that the major artery in the upper arm is the *brachial artery*. It is a continuation from above of
axillary	the _____ artery.
	46 The axillary artery itself is a continuation of the
subclavian	_____ artery.
innominate	**47** The subclavian artery is a branch of the _____ artery.
	48 The brachial artery gives two branches that enclose the head of the humerus. They are the anterior and the posterior
humeral	circumflex _____ arteries.
distal	**49** The brachial artery usually bifurcates just (distal/ proximal/lateral) to the elbow joint.
	50 The two branches into which it bifurcates are the
ulnar; radial	_____ and _____ arteries.
	51 The radial artery is so named because it lies principally on
radial	the _____ bone of the forearm.
	52 The blood supply to the palm of the hand is via the
palmar	superficial and deep _____ arches.

34

4 The forearm and hand

1 The forearm contains many more muscles than the upper arm, although their total bulk is less. The muscles originate both on the epicondyles of the humerus and on the

radial; ulnar

_____ bone and the _____ bone of the forearm.

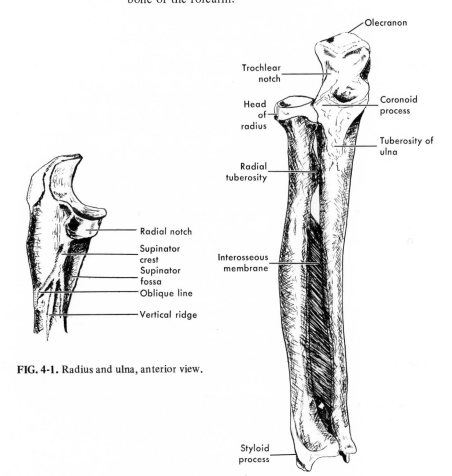

Radial notch
Supinator crest
Supinator fossa
Oblique line
Vertical ridge

Olecranon
Trochlear notch
Head of radius
Coronoid process
Tuberosity of ulna
Radial tuberosity
Interosseous membrane
Styloid process

FIG. 4-1. Radius and ulna, anterior view.

2 Examine Fig. 4-1 and note that the olecranon onto which

35

triceps	the _____ muscle inserts is a projection of the ulna.
distal	3 The styloid process of the radius is at the (proximal/distal) end of the radius.
	4 The fibers of the interosseous membrane run upward from
ulna to radius	(radius to ulna/ulna to radius).
proximal	5 The radius has a tuberosity at its (proximal/distal) end onto
biceps	which the _____ muscle of the upper arm inserts.
	6 There are two processes at the proximal end of the ulna, the
olecranon	_____ process on the posterior
coronoid	side and the _____ process on the anterior side.
	7 At the wrist the radius alone articulates with the *carpal*

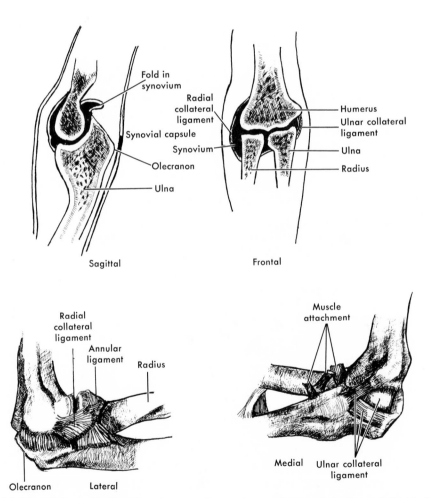

FIG. 4-2. Elbow joint. In the frontal section note the prominence of the medial epicondyle (right elbow).

36

bones. However, the ulna at its proximal end has a more

humerus

extensive articulation with the _____ than does the radius.

proximal

8 The supinator crest and oblique line are on the (proximal/

ulna

distal) end of the (ulna/radius).

9 Examine Fig. 4-2 of the elbow joint and note the three

radial collateral

major ligaments: the _____

medial collateral

ligament on the lateral side, the _____

_____ ligament on the medial side, and the

annular

_____ ligament holding the head of the radius.

10 The ligament that holds the head of the radius to the ulna, a

annular

ringlike ligament called the _____ ligament, is tapered from distal to proximal so that the head of the radius cannot be pulled forward or distally.

11 Note the articular capsule of the elbow. It has a large

fully extended

posterior fold so that the joint may be (fully extended/fully flexed/rotated).

12 The elbow joint itself is a hinge joint. Pronation and supination of the hand therefore take place elsewhere. Palpate your forearm while pronating and supinating your

radioulnar joint

hand. The action takes place at the (shoulder/wrist/ radioulnar joint).

ORIGINS OF THE MUSCLES OF THE FOREARM

13 As in the upper arm, the muscles of the forearm are grouped

anterior

into two compartments: the flexors in the (anterior/

posterior

posterior) compartment and the extensors in the (anterior/ posterior) compartment.

14 The muscles of the forearm arise principally from the medial

humerus

and lateral epicondyles of the _____.

15 Supinate your right hand. Palpate the muscles on the anterior surface near the elbow while you flex and extend

medial

your wrist. The wrist flexors are attached to the (medial/ lateral) epicondyle.

16 The muscles on the anterior surface of the forearm

flex

(flex/extend) the elbow or wrist or fingers and are divided into *five* superficial muscles and *three* deep ones.

five

17 The names of the (five/three/four) superficial muscles of the flexor group of the forearm are:

> pronator teres
> flexor carpi radialis
> palmaris longus
> flexor carpi ulnaris
> flexor digitorum superficialis

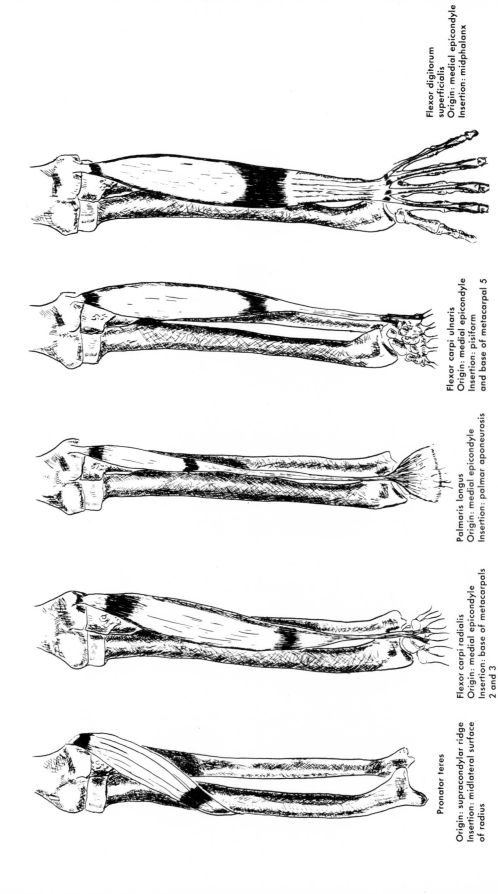

Pronator teres

Origin: supracondylar ridge
Insertion: midlateral surface
of radius

Flexor carpi radialis
Origin: medial epicondyle
Insertion: base of metacarpals
2 and 3

Palmaris longus
Origin: medial epicondyle
Insertion: palmar aponeurosis

Flexor carpi ulnaris
Origin: medial epicondyle
Insertion: pisiform
and base of metacarpal 5

**Flexor digitorum
superficialis**
Origin: medial epicondyle
Insertion: midphalanx

FIG. 4-3. The origin of the superficial muscles of the anterior compartment of the forearm is the medial epicondyle, with the exception of the pronator teres.

A mnemonic for these muscles is "planted flowers produce foul floods."

18 The pronator teres is the only muscle not arising from the medial epicondyle of the humerus. The pronator teres arises from the supracondylar ridge of the humerus, whereas the

medial epicondyle

remaining four muscles arise from the _____

_____ of the humerus.

19 The flexor carpi (= wrist) radialis indicates by its name that it

radial

runs along the _____ bone of the forearm.

20 The palmaris longus is often missing and is a weak and unimportant muscle. From its name one could guess that it

palm

inserts into the _____ of the hand.

21 The flexor carpi ulnaris is a counterpart to the flexor carpi

medial

radialis and therefore will be found on the (medial/lateral) side of the forearm.

22 The flexor digitorum (= finger) superficialis obviously must

fingers

insert onto the (wrist/palm/fingers).

four

23 Examine Fig. 4-3 and note that there are (three/five/four) muscles that are multijoint muscles (two joints or more).

flexes

24 Pronator teres performs two functions: it (flexes/extends)

pronates

the elbow and _____ the hand as its name suggests.

flexion

25 Flexor carpi radialis assists in (flexion/extension) of the

flexor

elbow and is a primary (extensor/flexor) of the wrist.

26 Palmaris longus, although thin and weak or in some cases

flex

missing altogether, would also act to _____ the elbow and the wrist.

27 Flexor carpi ulnaris is the counterpart of the flexor radialis

flex

and acts to _____ the elbow and to flex the

wrist

_____ . Note in Fig. 4-4 the relationships of the radialis and ulnaris.

28 Flexor digitorum, judging from its very extensive insertion,

fingers

is primarily a flexor of the _____ and

wrist; elbow

secondarily a flexor of the _____ and _____ .

29 From a study of their insertions, one can assume also that the flexor carpi ulnaris and flexor carpi radialis have yet another function. Flexor carpi radialis also assists in abducting the hand, and flexor carpi ulnaris also assists in

adducting

_____ the hand.

30 In the action of hammering (abducting and adducting the

flexor carpi radialis

hand) a carpenter uses the _____

flexor carpi ulnaris

and _____ muscles.

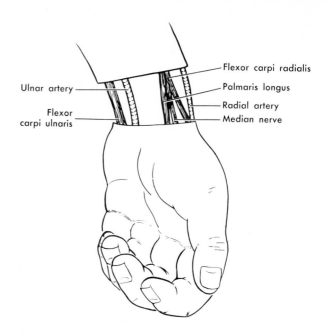

Ulnar artery

Flexor carpi ulnaris

Flexor carpi radialis

Palmaris longus

Radial artery

Median nerve

FIG. 4-4. Relationship of the tendons, arteries, and nerves passing through the wrist. The palmaris longus may be absent in one or both wrists in some people.

31 All the muscles of the superficial group are innervated by the *median* nerve, with the exception of the flexor carpi ulnaris, which is supplied by the *ulnar* nerve. Thus in the action of "curling" a barbell, a weight trainer would use the

flexor digitorum super-
ficialis

median

_____ muscle to flex his

fingers. This muscle is innervated by the _____ nerve.

32 He would use the remaining muscles of the anterior compartment to flex his wrist and assist in flexing his elbow. The remaining muscles would be stimulated by the

median; flexor
carpi ulnaris

_____ nerve except for the _____

_____ muscle, which is stimulated by the ulnar nerve.

33 A study of the surface anatomy of the tendons at the wrist will show that flexing the wrist against resistance will make two or perhaps three tendons stand out. On the right wrist from medial to lateral they are the flexor carpi ulnaris, palmaris longus, and the flexor carpi radialis. Judging from

flexor carpi ulnaris

flexor carpi
radialis

palmaris longus

their insertions, one can say that the _____ muscle has the most medial tendon, the _____

_____ has the most lateral tendon, and the

_____ is median (median=central).

40

34 The tendons are guides to the location of underlying nerves and blood vessels. The palmaris longus lies superficial to the

median

nerve that innervates it, the _____ nerve.

35 The flexor carpi radialis is a guide to the location of an

radial

artery at the wrist, the (ulnar/radial) artery. Examine Fig. 4-4 and note the relationships.

36 The flexor carpi ulnaris tendon is a guide to both an artery

ulnar; ulnar

and a nerve, the (radial/ulnar) artery and the (ulnar/radial) nerve.

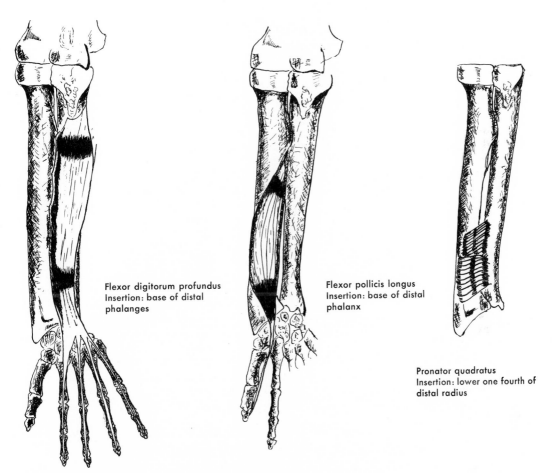

Flexor digitorum profundus
Insertion: base of distal
phalanges

Flexor pollicis longus
Insertion: base of distal
phalanx

Pronator quadratus
Insertion: lower one fourth of
distal radius

FIG. 4-5. Three deep muscles of the anterior compartment of the forearm. The pronator quadratus is the deepest muscle of the compartment.

37 The deep muscles of the forearm (Fig. 4-5) originate on the radius and ulna. The muscles are:

flexor digitorum profundus
flexor pollicis longus
pronator quadratus

41

fingers	The first arising is primarily a muscle of the (wrist/elbow/fingers).
	38 The flexor digitorum profundus (= deep) originates on the ulna and interosseous membrane, whereas the flexor pollicis (= thumb) longus originates on the other bone of the
radius	forearm, the _____ , and the interosseous membrane.
	39 The pronator quadratus also originates from the ulna, but, unlike the pronator teres, the pronator quadratus is situated
distal	at the (distal/proximal) end of the forearm.
two	**40** Of the deep muscles, (one/two/three) arise(s) in part from the ulna.
	41 The interosseous membrane serves to join the radius to the
origin	ulna. It also is a site of (origin/insertion) for certain muscles of the forearm.
	42 From their names, origins, and insertions, one can ascertain the functions of each muscle of the deep group of flexors.
flexes	The flexor digitorum profundus (flexes/extends) the distal *phalanx* (Fig. 4-10) of each finger on the medial phalanx.
	43 The flexor pollicis longus performs the action of
flexion	_____ on the thumb.
pronation	**44** The pronator quadratus assists in _____ of the hand.
	45 The deep muscles must be arranged in layers to reach their
pronator quadratus	insertions. The deepest of the three is the _____ _____ .
	46 The innervation of the muscles is by a branch of the main
median	nerve of the anterior compartment, the _____ nerve. This nerve branch, the *anterior interosseous* nerve, stimulates all the muscles; however, the ulnar nerve excites the medial half of the flexor digitorum profundus as well as
flexor carpi ulnaris	the _____ muscle of the superficial group.
two	**47** The ulnar nerve sends branches to (one/two/three) muscles of the flexors of the wrist and fingers (both superficial and deep groups).
	48 The muscles excited by the ulnar nerve are the
flexor carpi ulnaris; flexor digitorum profundus	_____ and the _____ .
	49 The blood supply to the muscles of the flexor compartment
radial	is via two arteries, the _____ and
ulnar	_____ , which arise from the major

brachial	artery of the upper arm, the _____ .

POSTERIOR COMPARTMENT OF THE FOREARM

	50 The muscles of the extensor group of the wrist and hand,
two	like the flexors, are divided into (three/two/four) groups.
	51 There are seven superficial and five deep extensor muscles.
four	There are thus _____ more extensor muscles than flexor muscles.
seven	**52** The (five/six/seven) superficial muscles (Fig. 4-6) are:

brachioradialis	extensor digitorum
extensor carpi radialis longus	extensor digiti minimi
extensor carpi radialis brevis	anconeus
extensor carpi ulnaris	

The mnemonic for these muscles is "blue-clawed crabs undulate daintily down avenues."

53 The seven superficial extensor muscles originate on the lateral epicondyle of the humerus except for the brachioradialis and the extensor carpi radialis longus. These latter two muscles originate above the lateral epicondyle on the lateral

supracondylar	_____ ridge of the humerus just as
supracondylar	the pronator teres originated on the medial _____ ridge.
supracondylar	**54** The brachioradialis arising on the _____ ridge of the humerus and inserting onto the styloid process of the radius is unusual in that it is actually a flexor, *not* an extensor! It is included with the superficial extensors because it is innervated by the radial nerve that innervates all the muscles of this group. The brachioradialis flexes the
elbow	_____ joint.
	55 The extensor carpi radialis longus and brevis are obviously counterparts to one muscle of the superficial flexor group,
flexor carpi radialis	the _____ .
lateral	**56** The two extensor muscles are found on the (lateral/medial) aspect of the forearm.
	57 The extensor carpi ulnaris also has a counterpart in the
flexor carpi ulnaris	flexor group, the _____ .
	58 If one were to think of the wrist as a square, there would be muscles attached at all four corners. The lateral two corners
flexor carpi radi- alis; extensor carpi radialis longus; exten- sor carpi radialis brevis	would have three muscles attaching: the _____ _____ , the _____ _____ , and the _____ _____ .

flexor carpi ulnaris

extensor carpi ulnaris

extensor digitorum su-
perficialis; extensor
digitorum profundus

little finger

59 The two muscles attaching to the medial two corners of the wrist would be the _____ and the _____.

60 The extensor digitorum obviously is the antagonist to two muscles of the flexor surface, the _____ _____ and the _____ _____.

61 The extensor digiti minimi is an extra muscle of the (thumb/little finger/middle finger).

62 The anconeus is an extensor muscle of the elbow and arises on the back of the lateral epicondyle. It is another

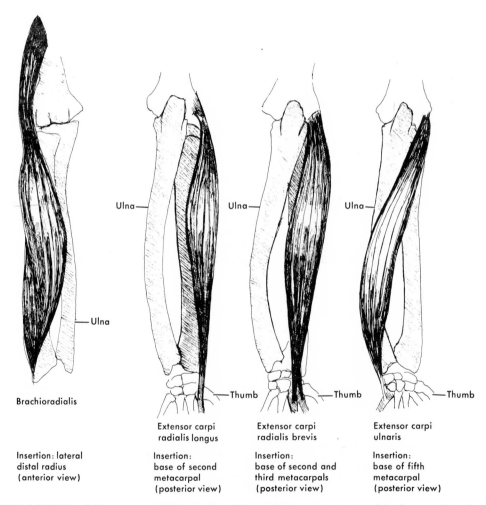

Brachioradialis

Insertion: lateral
distal radius
(anterior view)

Extensor carpi
radialis longus

Insertion:
base of second
metacarpal
(posterior view)

Extensor carpi
radialis brevis

Insertion:
base of second and
third metacarpals
(posterior view)

Extensor carpi
ulnaris

Insertion:
base of fifth
metacarpal
(posterior view)

FIG. 4-6. Origin of the seven superficial muscles of the posterior compartment of the forearm. Note the common origin is the lateral epicondyle with a few exceptions.

brachioradialis

interloper, in this group, just as the _____
muscle is. The anconeus is included with this group because
it is innervated by the same nerve as the others, the

radial

_____ nerve.

five

63 Examine Fig. 4-6. Note that there are (seven/six/five/three)
muscles that are multijoint muscles (two joints or more).

64 Note also that all but three of the muscles arise from a
common tendon originating on the lateral epicondyle of the

brachioradialis
extensor carpi radialis
longus
lateral supracondylar
ridge

humerus. The three exceptions are the _____

and _____ ,

arising on the _____ ,

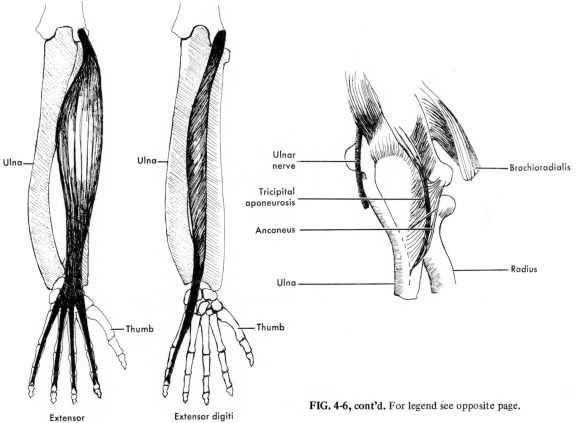

Ulna

Ulna

Ulnar
nerve

Tricipital
aponeurosis

Anconeus

Ulna

Brachioradialis

Radius

Thumb

Thumb

Extensor
digitorum

Insertion:
extensor expansion
fingers 2, 3, 4, 5

Extensor digiti
minimi

Insertion: with
extensor digitorum
to extensor expansion
finger 5

FIG. 4-6, cont'd. For legend see opposite page.

45

anconeus

lateral epicondyle

extend

flexor carpi radialis

flexor carpi ulnaris

abduction-adduction

extensor digitorum

extensor digiti minimi

extended

extensor

radial

two

extension

five

supinator

lateral epicondyle

ulnar

and the _____, arising on the back of the _____.

65 As with the flexor muscles, one can ascertain the functions of the extensors largely by their names. The extensor carpi radialis longus and brevis as well as the extensor carpi ulnaris act to (flex/extend) the wrist and hand.

66 The same muscles also act with their counterparts in the flexor group, the _____ and the _____ , to help produce actions such as hammering that combine the actions of (flexion-extension/abduction-adduction/pronation-supination).

67 The extensors of the fingers themselves are the _____ and the _____ muscles.

68 When preparing to cast in fly fishing, the fisherman draws the rod backward over his head. The wrist is cocked in an _____ position produced by the _____ muscles of the wrist, which are innervated by the _____ nerve.

69 The fifth digit is different from the other digits in that it has (three/one/two) superficial extensor muscles attached to it.

70 The anconeus muscle plays no part in extension of the wrist and fingers. Its job is (pronation/flexion/extension) of the elbow.

71 There are (five/three/four) deep extensors of the forearm.

72 The deep extensors from proximal to distal are:

supinator
abductor pollicis longus
extensor pollicis longus
extensor pollicis brevis
extensor indicis

A mnemonic for remembering the group is "sailors always enlist by inches."

73 There is no common origin for these muscles. The most proximal muscle, the _____ , arises in part from the common point of origin of the superficial extensor muscles, the _____ .

74 The supinator also arises from the supinator crest and oblique line of the _____ bone of the forearm.

75 The abductor pollicis longus originates on the

46

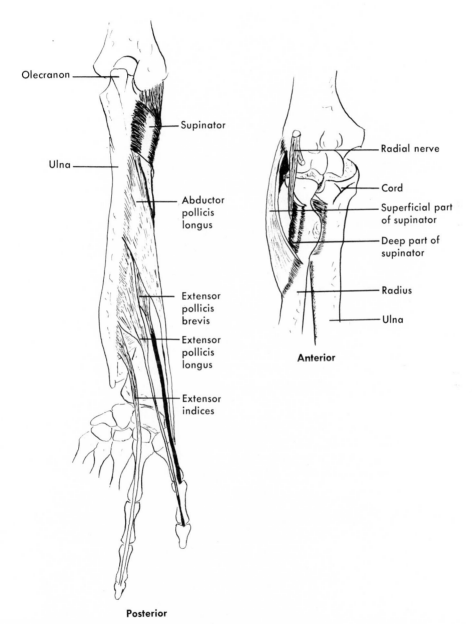

Posterior

Anterior

Labels (left/posterior view): Olecranon, Supinator, Ulna, Abductor pollicis longus, Extensor pollicis brevis, Extensor pollicis longus, Extensor indices

Labels (right/anterior view): Radial nerve, Cord, Superficial part of supinator, Deep part of supinator, Radius, Ulna

FIG. 4-7. Origin of the five deep muscles of the posterior compartment. Note that the supinator has two portions; the deeper of the two wraps around the shaft of the radius from behind. The tendons of the three pollicis muscles form the anatomical snuffbox.

interosseous	_____ membrane joining the radius and ulna and on the adjacent surfaces of the radius and ulna.
smaller	76 The extensor pollicis longus originates on the interosseous membrane and the ulna. It therefore has a (larger/smaller) area of origin than the abductor pollicis longus.
extensor pollicis brevis	77 The next muscle in order is the _____ , and it originates on the interosseous membrane and the radius.
extensor indicis	78 The last muscle, the _____ originates on the interosseous membrane and the distal part of the ulna.
three	79 An examination of the deep extensors (Fig. 4-7) reveals that (four/two/three) are muscles of the thumb.
supinator; palm up	80 The function of the first muscle of the deep extensors, the _____ , is turning the hand to a (palm up/palm down) position.
ulna	81 The supinator runs from its origins on the _____ to the radius near the tuberosity, wrapping around the back of the radius to get to its insertion.
abduct	82 The abductor pollicis longus must function to _____ the thumb.
supinated	83 The abduction of the thumb is somewhat different from abduction of other digits. Place your hand knuckles down on a table, and point your thumb toward the ceiling. The thumb in this position is abducted, and the hand is (pronated/supinated).
extension	84 The extensor pollicis longus is responsible for _____ of the thumb. This is demonstrated by holding the hand pronated on a flat surface and pulling the thumb away from the hand.
extensor pollicis longus	85 The extensor pollicis brevis assists the _____ in extending the thumb.
	86 The tendons of these latter three muscles give rise to the borders of a triangular area known as the *anatomical snuffbox* (Fig. 4-8). The muscle tendons at the thumb from lateral to medial are:
abductor pollicis longus	_____
extensor pollicis brevis	_____
extensor pollicis longus	_____
	87 The anatomical snuffbox can be demonstrated by holding your right hand in pronation and extending the thumb as far as possible and pointing it toward you.

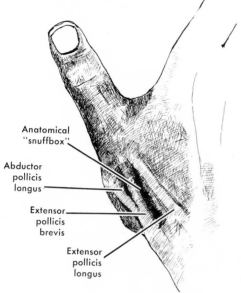

FIG. 4-8. Anatomical snuffbox outlined by the tendons of the extrinsic muscles of the thumb in the posterior compartment.

Anatomical
"snuffbox"

Abductor
pollicis
longus

Extensor
pollicis
brevis

Extensor
pollicis
longus

radial

supinator

ulnar

carpal

eight

capitate

88 The muscles of the deep extensors are innervated by branches from the major nerve of the extensors, the

_____ nerve. The most important branch is the *posterior interosseous* nerve.

89 In the action of making a mock gun with the thumb and forefinger (with the wrist extended) nearly all the deep extensors of the forearm are involved except the (extensor indicis/supinator/abductor pollicis longus).

90 The blood supply to the extensor muscles is largely carried by the *posterior interosseous* artery, which lies between the two groups of muscles. It is a branch of the (axillary/brachial/ulnar) artery (Fig. 3-5).

91 The hand includes the wrist, or carpal bones. The carpal bones begin at the most distal horizontal fold of skin of the

flexor surface of the wrist. The wrist, or _____ , bones are eight in number.

92 The (six/eight/seven) carpal bones are arranged in two rows, a proximal row and a distal row.

93 Examine Fig. 4-9 and note that the names of the bones in the distal row from left to right are hamate, capitate, trapezoid, and trapezium. The bones in the proximal row from left to right are pisiform, triquetrum, lunate, and scaphoid.
The mnemonic for the carpals is "holy cow, Trixie and Tom pilfered a truck last Sunday."

94 The largest bone in the distal row is the _____ .

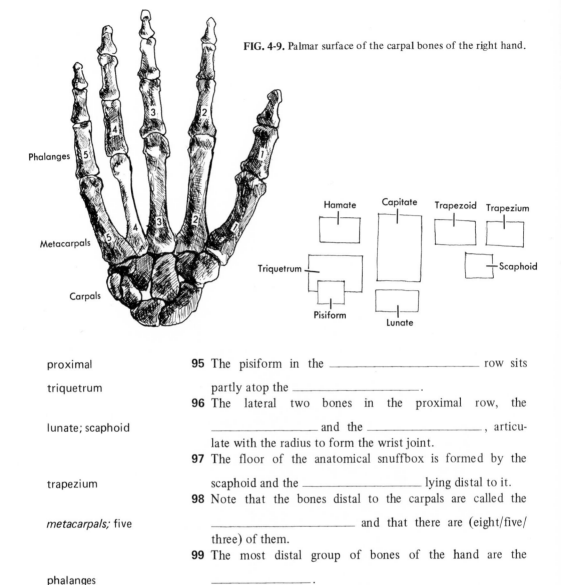

FIG. 4-9. Palmar surface of the carpal bones of the right hand.

proximal

triquetrum

lunate; scaphoid

trapezium

metacarpals; five

phalanges

three
first

five

flexor carpi ulnaris

radius

95 The pisiform in the _____ row sits partly atop the _____ .

96 The lateral two bones in the proximal row, the _____ and the _____ , articulate with the radius to form the wrist joint.

97 The floor of the anatomical snuffbox is formed by the scaphoid and the _____ lying distal to it.

98 Note that the bones distal to the carpals are called the _____ and that there are (eight/five/three) of them.

99 The most distal group of bones of the hand are the _____ .

100 Each digit has _____ phalanges except the (first/second/fifth).

INSERTIONS OF THE MUSCLES OF THE FOREARM

101 An examination of the insertions of the *flexors* of the wrist in Fig. 4-10 shows that of the (six/four/five) in the superficial group, only one inserts on a carpal bone. It is the _____ .

102 The pronator teres inserts on the lateral surface of the _____ .

50

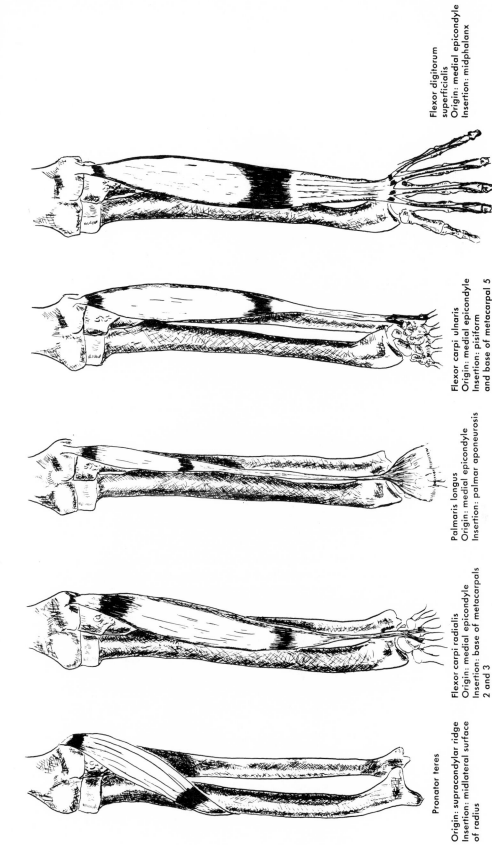

Pronator teres

Origin: supracondylar ridge
Insertion: midlateral surface
of radius

Flexor carpi radialis
Origin: medial epicondyle
Insertion: base of metacarpals
2 and 3

Palmaris longus
Origin: medial epicondyle
Insertion: palmar aponeurosis

Flexor carpi ulnaris
Origin: medial epicondyle
Insertion: pisiform
and base of metacarpal 5

Flexor digitorum
superficialis
Origin: medial epicondyle
Insertion: midphalanx

FIG. 4-10. Insertions of the flexors of the hand and fingers.

103 The flexor carpi ulnaris inserts on the pisiform and fifth metacarpal, whereas the flexor carpi radialis inserts on the (third and fourth/first and second/second and fourth) metacarpals.

first and second

104 The palmaris longus has a general insertion into the (aponeurosis of the palm/interosseous membrane).

aponeurosis of the palm

105 The flexor digitorum superficialis, since it flexes the fingers, must insert onto the (metacarpals/carpals/phalanges).

phalanges

106 Of the deep flexors, the flexor digitorum profundus also inserts on the phalanges but onto the most (distal/proximal) phalanges.

distal

107 The more distal insertion of the deeper muscle is made possible by the fact that the flexor digitorum superficialis

flexor digitorum
  profundus

splits to allow the deeper placed _____

_____ muscle to penetrate and thus reach more distally.

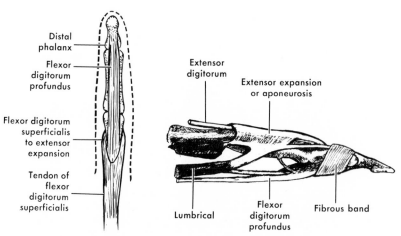

Distal phalanx

Flexor digitorum profundus

Flexor digitorum superficialis to extensor expansion

Tendon of flexor digitorum superficialis

Extensor digitorum

Extensor expansion or aponeurosis

Lumbrical

Flexor digitorum profundus

Fibrous band

Palmar aspect

FIG. 4-11. The flexor digitorum superficialis and profundus travel together. Note how the superficialis tendon divides to reach the extensor expansion and the profundus proceeds to the distal phalanx.

108 A wrapping of fibrous tissue on the extensor surface of the fingers known as the *extensor expansion* (Fig. 4-11) receives the insertions of several muscles, both intrinsic and extrinsic.

extensor expansion

The wrapping called the _____ on the dorsum (= back) of the fingers provides a generalized insertion for the flexor digitorum superficialis, and the profundus muscle inserts directly onto the (distal/proximal) phalanx.

distal

109 Examine Fig. 4-12 and note that none of the extensor muscles attaches to the back of a (carpal/metacarpal/phalange).

carpal

52

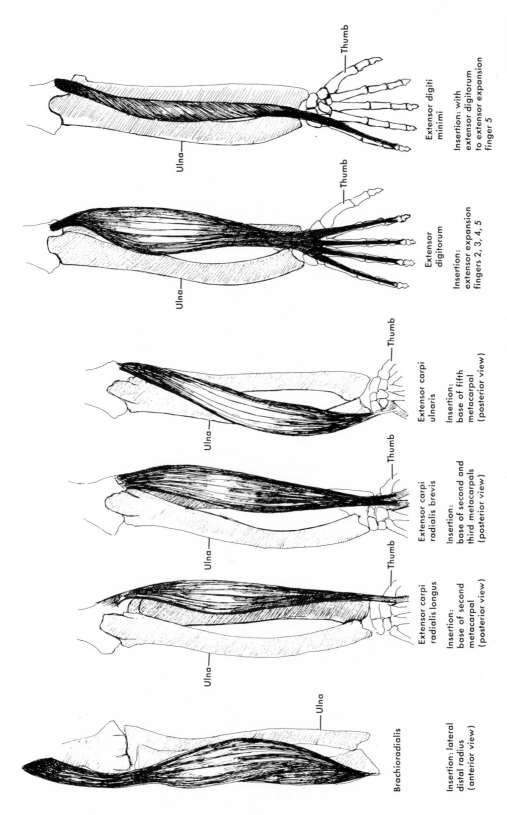

Brachioradialis

Insertion: lateral
distal radius
(anterior view)

Extensor carpi
radialis longus

Insertion:
base of second
metacarpal
(posterior view)

Extensor carpi
radialis brevis

Insertion:
base of second and
third metacarpals
(posterior view)

Extensor carpi
ulnaris

Insertion:
base of fifth
metacarpal
(posterior view)

Extensor
digitorum

Insertion:
extensor expansion
fingers 2, 3, 4, 5

Extensor digiti
minimi

Insertion: with
extensor digitorum
to extensor expansion
finger 5

FIG. 4-12. Insertions of the seven superficial muscles of the posterior compartment.

Continued.

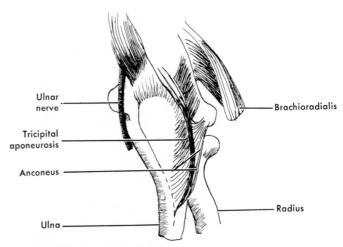

Ulnar nerve

Brachioradialis

Tricipital aponeurosis

Anconeus

Radius

Ulna

FIG. 4-12, cont'd. For legend see p. 53.

does not

110 The brachioradialis (does/does not) cross the wrist. It inserts

styloid

on the radius above the _____ process.

111 The extensor carpi radialis longus inserts on the base of the

metacarpal

second (carpal/metacarpal/phalange).

112 The extensor carpi radialis brevis inserts on the base of the

second and third

(second and third/first and second/first and third) metacarpal bones.

113 The position of the insertion of the *flexor* carpi radialis is

extensor carpi radialis
brevis

somewhat similar to the insertion of the (extensor carpi radialis longus/extensor carpi radialis brevis).

114 The extensor carpi ulnaris inserts on the opposite surface of the same metacarpal as the flexor carpi ulnaris, the

fifth metacarpal

_____ bone.

115 Since the extensor digitorum extends the fingers, it must

phalanges

insert on the (phalanges/carpals/metacarpals) via the extensor expansion.

116 Although there is no extensor digitorum profundus as there is a flexor digitorum profundus, there are extensor muscles to two fingers in addition to the extensor digitorum muscles already mentioned. These two extra extensor digiti muscles

extensor digiti minimi

in the deep layer are the _____

extensor indicis

and the _____ .

117 The additional extensors insert on the extensor expansion of

second; fifth

the _____ and _____ fingers.

118 The anconeus inserts on the olecranon and the oblique line

ulna

of the (radius/ulna/humerus).

54

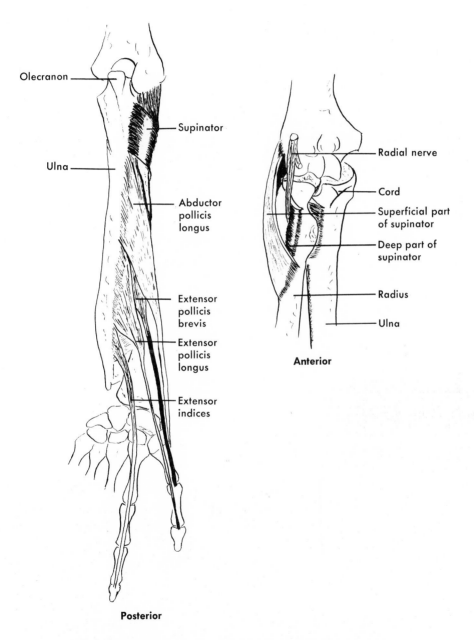

Olecranon

Supinator

Ulna

Abductor
pollicis
longus

Extensor
pollicis
brevis

Extensor
pollicis
longus

Extensor
indices

Posterior

Radial nerve

Cord

Superficial part
of supinator

Deep part of
supinator

Radius

Ulna

Anterior

FIG. 4-13. Insertions of the five deep muscles of the posterior compartment.

55

distal phalanx	**119** The flexor pollicis longus reaches as far as the (proximal phalanx/distal phalanx).
	120 A longus muscle implies that there must be a
brevis	_____ muscle.
five	**121** The deep extensors of the forearm (Fig. 4-13) are (five/three/four) in number.
	122 Apart from the supinator and extensor indicis, the deep
thumb	muscles all concern one digit, the _____ .
	123 The supinator has two parts, one of which "wraps" around
radius	the (ulna/radius/humerus) to provide the supinating action to the hand.
	124 The *extrinsic* (= external) extensor muscles of the thumb from proximal to distal on the extensor surface are
abductor pollicis longus	_____
extensor pollicis brevis	_____
extensor pollicis longus	_____
extensor pollicis longus	**125** Of the thumb muscles, the tendon of the _____
	_____ reaches the farthest distally on the thumb.
distal phalanx	**126** It inserts on the (proximal phalanx/distal phalanx/middle phalanx).
first metacarpal	**127** The extensor pollicis brevis inserts on the (first metacarpal/proximal phalange).
	128 The abductor pollicis longus inserts into the base of the first metacarpal, and it abducts the first metacarpal at the
carpometacarpal	(metacarpal-phalangeal joint/carpometacarpal joint).
	129 The total number of *extrinsic* muscles of the thumb is
four	(four/two/three/one).
one; two	**130** These include (one/two/three) flexors, (one/two/three) extensors, and (one/two/three) abductors.
one	

INTRINSIC MUSCLES OF THE HAND

131 At this point, a second set on muscles that acts on the hand will be examined. The muscles originating on the humerus, radius, and ulna are called the *extrinsic muscles of the hand;* therefore those originating on the bones of the hand proper

intrinsic   are called the _____ muscles.

132 Examine your own palm with the hand in a relaxed position and supported. Note that in the relaxed state the fingers are

semiflexed   normally (extended/semiflexed/abducted).

133 Also note the ball of the thumb. It is known as the *thenar eminence.* Opposite, note the ball of the little finger, the *hypothenar eminence* (the term *thenar* originally meant palm).

56

134 Take a pinch of skin from the dorsum, or back, of the hand. Now try to take a pinch of skin from the palm. The skin is
palm — attached more firmly to the (palm/dorsum) of the hand.

135 The palm is capable of absorbing shocks such as slapping a hard surface because it contains fibrofatty pads. The
deep — tendons, nerves, and blood vessels pass (superficial/deep) to the pads.

136 The flexor tendons at the wrist are held in place by a tough fascia called the *flexor retinaculum,* which is about 3 cm.
flexor — wide. The retinaculum prevents the (flexor/extensor) tendons from "bowing" like the string on a bow when the wrist is flexed.

137 Flex your wrist to its maximum and then extend it to its maximum. In which action, flexion or extension, does the
flexion — greater range of motion lie? _____

138 The palmar aponeurosis is a continuation of the flexor retinaculum, and it reaches to the webs of the fingers. The
palmaris longus — _____ muscle is inserted into the palmar aponeurosis.

distal
139 Note on Fig. 4-14 that the flexor retinaculum lies (distal/proximal) to the flexion creases of the skin of the wrist.

140 Note also that the retinaculum forms a roof over the concavity provided by the carpal bones. The whole is known as the *carpal tunnel.* The nerve(s) passing through the tunnel
ulnar and median nerves — are the _____ . The ar-
ulnar artery — tery(ies) passing through the tunnel is the _____ .

141 Some of the intrinsic muscles of the thumb are the brevis counterparts to the extrinsic longus muscles. Thus there
abductor — should be an _____ pollicis brevis and a
flexor — _____ pollicis brevis.

142 In addition, the thumb is supplied with an *opponens* muscle. This muscle plus the abductor and flexor pollicis brevis
thenar — muscles comprise the _____ eminence on the thumb side of the hand.

143 The thumb also has an adductor pollicis in addition to its
three — (three/four/two) other intrinsic muscles.
brevis
144 The abductor pollicis (longus/brevis) of the thenar eminence arises from the retinaculum and from the scaphoid and trapezium.

145 Supinate your own hand and abduct the thumb (p. 48). Palpation reveals that the abductor pollicis brevis inserts on the
proximal — bone of the (distal/proximal) phalanx.

146 The flexor pollicis brevis arises from approximately the same source as the abductor pollicis brevis. It does not have attachment at the scaphoid but does have origin at the

57

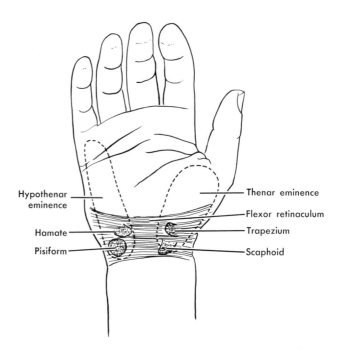

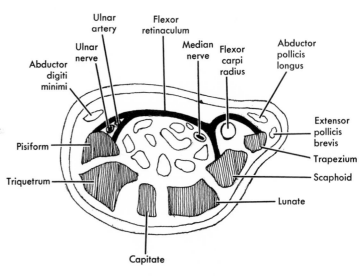

FIG. 4-14. The attachment of the flexor retinaculum provides a roof over the carpal tunnel through which most of the tendons, nerves, and blood vessels travel.

retinaculum; trapezium

_____and the _____ .

147 The flexor pollicis brevis inserts in common with the abductor pollicis brevis onto the base of the

proximal

_____ phalanx.

opponens

148 The third of the thenar muscles, the _____ ,

58

originates at the same site as the flexor pollicis, the

retinaculum; trapezium _____ , and the _____ .

149 The opponens muscle inserts onto the first metacarpal unlike the other two thenar muscles, which insert on the

first phalanx _____ .

150 Observe on Fig. 4-15 that the deepest thenar muscle is the

opponens _____ .

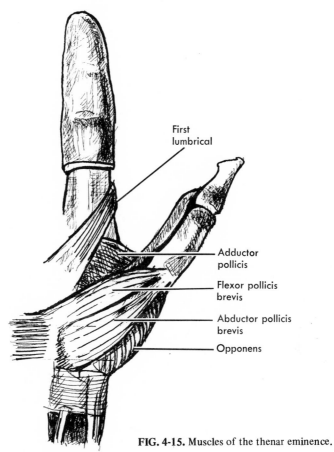

First
lumbrical

Adductor
pollicis

Flexor pollicis
brevis

Abductor pollicis
brevis

Opponens

FIG. 4-15. Muscles of the thenar eminence.

abductor pollicis **151** The thenar muscle most proximal is the _____
brevis
_____ .

four **152** There are thus (three/five/six/four) intrinsic muscles of the thumb.

three **153** Of the four, only (two/four/three) comprise the muscles of the thenar eminence.

154 The third named muscle of the thenar group is the deepest of the group, and its action is as its name suggests. It

opposing performs the act of _____ .

155 The opponens causes the thumb to work in opposition to the remaining fingers. Thus the thumb is brought to a position so that its tip can meet the tip of any other finger. Try touching the tip of your fifth finger to the thumb without moving your thumb.

156 The thenar muscles are stimulated by a branch

median _____ nerve that excites the flexor group of the forearm, the *recurrent* branch. The ulnar nerve stimulates nearly all the remaining intrinsic muscles of the hand.

157 Many actions of the thumb are due to the muscles of the thenar eminence and the uniqueness of the thumb construction, which is somewhat different from the other digits. Opposition and abduction have already been explained.

abductor pollicis longus Abduction is caused by two muscles, the _____

abductor pollicis brevis _____ and the _____

_____ .

median 158 The abductor brevis innervated by the _____ nerve.

159 The movement of flexion is also different from flexion in other digits.

supinated Flexion: with the hand held palm up, or (pronated/supinated), scratch your thumbnail across your palm toward the base of the fifth digit. The opposite movement is

extension _____ .

160 The adductor pollicis has two heads; both, like the remaining intrinsic muscles of the hand, are innervated by the

ulnar (radial/median/ulnar) nerve.

161 The oblique head of the adductor pollicis originates on the second metacarpal and two carpal bones, the trapezoid and

distal capitate of the (proximal/distal) row of the carpals.

162 The oblique head of the adductor pollicis originating on the

second metacarpal; trapezoid; capitate _____ , _____ ,

and _____ bones inserts on the medial aspect of the proximal phalanx of the thumb.

163 The transverse head of the adductor pollicis originates on the third metacarpal and has a common insertion with the oblique head. Thus the transverse head proceeds from its

third origin on the _____ metacarpal to its insertion on

proximal phalanx the _____ of the thumb.

164 The adductor pollicis would be involved in the action of

gripping firmly (flipping coins/gripping firmly).

165 A gymnast working on the horizontal bar would rely heavily on his adductor pollicis to maintain his grip. The muscle is

ulnar innervated by the (median/ulnar/radial) nerve.

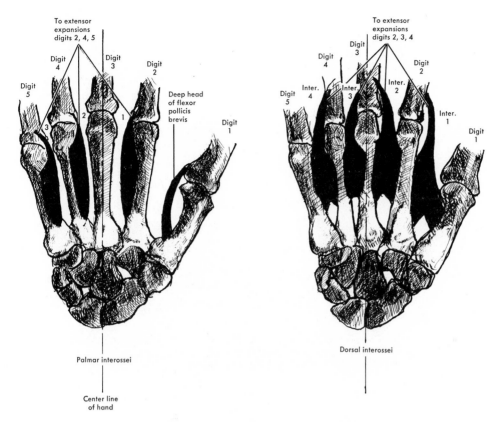

FIG. 4-16. *Left,* palmar interossei. *Right,* dorsal interossei. Note the direction the muscle runs in each case.

166 Note on Fig. 4-16 that the first dorsal interosseous muscle arises from the first metacarpal and the adjacent second metacarpal. It inserts with other muscles going to the dorsum

extensor

of the fingers on the _____ expansion of the index finger. Abduct your index finger against resistance, that is, spread it from the third finger, and you can easily see and feel the first dorsal interosseous muscle between the carpal of the thumb and the carpal of the index finger (dorsal surface).

167 The second dorsal interosseous muscle like the first arises from the adjacent second and third metacarpals but inserts

extensor expansion

into the _____ of the third, or middle, finger.

168 The third dorsal interosseous muscle arises from the adjacent

third and fourth

_____ metacarpals and inserts also into the extensor expansion of the middle finger.

second and third

169 The middle finger thus has the _____ dorsal interossei inserted onto its extensor expansion.

170 The fourth and last dorsal interosseous muscle arises from

61

the adjacent surfaces of the last two metacarpals, that is, the

fourth and fifth

_____ metacarpals, and inserts onto the extensor expansion of the ring finger.

171 From the position of the muscles in Fig. 4-16 you could deduce that the function of the dorsal interossei would be

abduction; from

(abduction/adduction) of the fingers (from/to) the midline of the hand and running through the middle finger.

172 The palmar interossei, of course, supply the action of

adduction

_____ .

173 The difference in the origins of the dorsal compared to the palmar interossei is that each dorsal interossei arises from

two bones

(ligaments/one bone/two bones).

174 All the palmar interossei muscles insert into the extensor expansions of the digit from whose metacarpal they (arise/

arise

are adjacent to).

175 The mnemonics to best remember the functions of the interossei are PAD (palmar adduct) and DAB (dorsal abduct).

176 The dorsal interossei are inserted into the extensor expan-

second, third, and
    fourth fingers

sions of (all fingers/the second, third, and fourth fingers/the first, second, and fourth fingers).

177 Note that the thumb and little finger have their separate abductors in the thenar and hypothenar eminences and

dorsal

therefore require no (palmar/dorsal) interossei.

178 The interossei have another function, that of *flexion*. They flex the metacarpal phalangeal joint (M.P. joint). Because of their unique insertion into the extensor expansion, they also serve to *extend* the distal and middle phalanges at the interphalangeal joints (I.P. joints). Their function in these capacities is best illustrated by the action of an infant waving good-bye with his fingers extended, and all movement taking place at the M.P. and I.P. joints.

179 The movement of waving, accomplished by the interossei

M.P.

and other muscles that flex the (M.P./I.P.) joint and extend

I.P.

the (M.P./I.P.) joints is assisted by yet another set of intrinsic muscle called the *lumbricals* (= wormlike).

three

180 There are (one/two/three/four) sets of intrinsic muscles acting on the second to fifth digits.

dorsal interossei

181 The three sets of muscles are _____

palmar interossei

_____ , _____ ,

lumbricals

and _____ .

182 Note on Fig. 4-17 that the four lumbrical muscles originate from the four tendons of the flexor digitorum profundus, and, like the other intrinsic muscles of the second to fifth

extensor expansions

digits, the lumbricals insert into the _____

_____ .

62

To extensor
expansions of
digits 2, 3, 4, 5

Tendons of
flexor digitorum
profundus

FIG. 4-17. Lumbricals. All insert from the radial side.

first

radial

ulnar

hypothenar

abductor digiti minimi
flexor digiti minimi; op-
ponens digiti minimi

183 Note on Fig. 4-15 that the (first/second/middle) two lumbricals originate on one tendon only.

184 Note also that all the lumbricals insert on the (radial/ulnar) side of the extensor expansion.

185 The first two lumbricals are innervated by the median nerve, whereas the second two are innervated, as are the interossei, by the _____ nerve.

186 The little, or fifth, finger has a set of three muscles on the palmar aspect known as the _____ eminence.

187 The three muscles (Fig. 4-18) are named as are their counterparts in the thenar eminence: _____ , _____ , and _____ .

188 Their functions are similar to those of the muscles of the thenar eminence except that the opponens draws the fifth metacarpal forward to deepen the hollow of the hand.

63

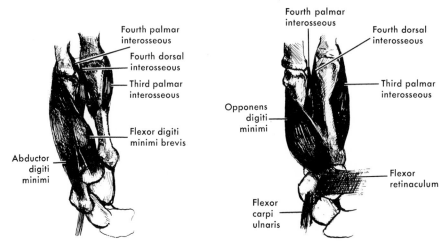

FIG. 4-18. Muscles of the hypothenar eminence. Note that they are named in a fashion similar to the muscles of the thenar eminence.

189 The muscle of the hypothenar eminence lying deepest is the

opponens

_____.

190 The muscles attach as follows:

abductor digiti minimi—origin: pisiform
        insertion: proximal phalanx, medial aspect
flexor digiti minimi—origin: hook of hamate
        insertion: proximal phalanx, medial aspect
opponens digiti minimi—origin: hook of hamate
        insertion: fifth metacarpal, medial aspect

191 The abductor digiti minimi serves the same purpose for the
fifth digit as the (palmar/dorsal) interosseous (Fig. 4-16)
does for the fourth digit.

dorsal

192 The abductor digiti minimi originates on the pisiform bone
of the (proximal/distal) row of carpals.

proximal

193 From its origin on the _____, the abductor
digiti minimi inserts onto the proximal phalanx of the little
finger.

pisiform

194 The flexor digiti minimi inserts in common with the
abductor onto the _____ of the
little finger, but it arises on the hamate (actually on the
hook of the hamate) bone of the (proximal/distal) row of
carpals.

proximal phalanx

distal

195 The opponens originates on the same site as the flexor digiti
minimi, that is, the _____ bone, and
then inserts onto the fifth metacarpal.

hamate

196 Note that two muscles of the hypothenar eminence originate
on the same bone, the _____ bone.

hamate

197 Note also that two have a common insertion on the

abductor digiti minimi

proximal phalanx, the _____

flexor digiti minimi

and the _____ .

198 The intrinsic muscles of the hand give accurate and precise movements, whereas the extrinsic muscles give power. The

intrinsic

action of buttoning a coat utilizes primarily the (extrinsic/ intrinsic) muscles.

power
extrinsic

199 The act of gripping a tennis racket requires (precision/power) and therefore the muscles of the (intrinsic/extrinsic) group.

200 The upstroke in writing requires the intrinsic muscles, which operate the fingers to extend the I.P. joint and flex the M.P.

lumbricals
palmar interossei
dorsal interossei

joint. The three groups of muscles are _____ ,

_____ , and _____ .

201 Make a tight fist and then flex your wrist as far as possible.

open slightly

Note that on extreme flexion of the wrist the fingers (close tighter/open slightly).

action of the opposite-
    pulling extensor
    muscles

202 Your fingers were pulled open by the (action of the opposite-pulling extensor muscles/inability of the flexor retinaculum to hold the tendons in).

$\mathbb{5}$ The hip

The lower limb has many analogies to the upper limb. There are, for example, two bones in the forearm and two bones in the leg (the leg is that portion from knee to ankle) and tarsal *and* metatarsal *bones that compare to the carpal and metacarpal bones of the wrist and hand. The hip joint is a ball-and-socket joint, as is the shoulder. There are differences too, of course, between the limbs. Most of those differences are concerned with the different functions of the upper and lower limbs.*

1 The hip joint, although tighter and stronger than the shoulder joint, is similar in many respects to the shoulder joint, just as that area is termed the shoulder girdle, the area of the hip is

girdle

called the *pelvic* _____ .

ball-and-socket

2 The hip and shoulder joints are both (hinge/ball-and-socket) joints.

3 The bones of the pelvic girdle are fused, and the ball-and-socket hip joint is much more complete. This makes the hip

more

(more/less) stable than the shoulder.

all are correct to some degree

4 The hip joint must be stable because it (carries the body weight/does not have to perform so many functions/is a by-product of standing erect).

5 Each hip is made up of three bones fused together:

> pubis
> ischium
> ilium

These three bones comprising the hip are fused into one at the *acetabulum,* or socket, for the hip joint.

ilium

6 The right pelvic girdle composed of the pubis, the _____,

ischium

and the _____ is joined in front to the left pelvic girdle at the *pubis symphysis* and in the rear at the *sacroiliac* joint.

7 The *sacrum,* a part of the spine, is a wedge of bone between the left and right ilium that helps form the immoveable joint

sacroiliac

at the rear of the pelvis, the _____ joint.

66

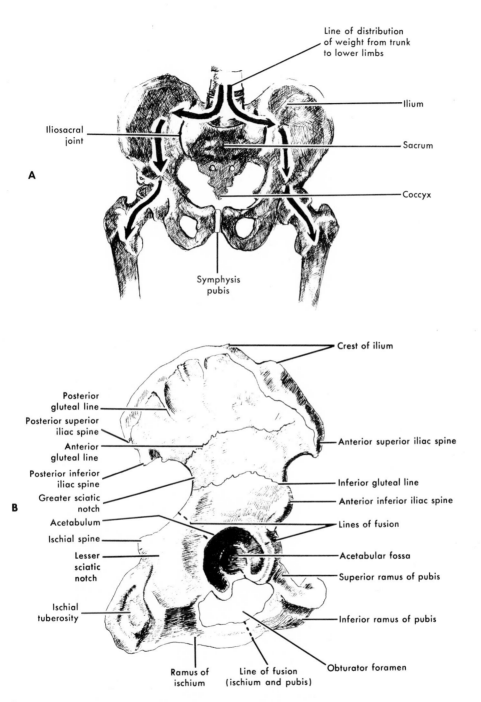

FIG. 5-1. **A,** The pelvis distributes the body weight to the supporting columns of the legs. **B,** Lateral view of the pelvis. Note the fusion at the acetabulum of the three bones—the ilium, ischium, and pelvis.

8 Examine Fig. 5-1 and note that the structure at the hip

acetabulum

analogous to the glenoid fossa is the _____ .

9 Palpate the crest of your own ilium and also locate the *anterior superior iliac spine* on your ilium. The latter is the site of a fairly common football injury called a *hip-pointer*

anterior superior
iliac spine

injury. The site of hip-pointer injury, the _____

_____ is an

important site of muscle attachment.

10 The *ischial tuberosity* is a thickened area on the posterior por-

ischium

tion of the _____ . From its position, you could deduce that it is important for weight bearing in the

seated
inferior

(standing/seated) position. The ischial spine marks the (inferior/superior) limit of the greater sciatic notch.

11 The ischial tuberosity is that portion of the skeleton in direct contact with the chair when one is seated. It is overlaid

inferior

somewhat with a fat pad and is (inferior/superior) to the acetabulum.

12 The pelvis has a large foramen on each side called the *obturator foramen,* which is normally closed by a membrane.

obturator

The _____ foramen transmits several important nerves and blood vessels to the lower limb.

13 The pelvic bone that comes closest to paralleling the structure

pubis

and function of the clavicle is the_____ .

14 The pelvic bone that because of its broad ala (ala = wing)

ilium

most resembles the scapula is the _____ .

ilium

15 There are faint lines on the ala of the _____ that mark the attachment of three *gluteal* muscles.

16 The three muscles arising on the ilium called the

gluteal

_____ *muscles* insert on the thigh bone, or femur.

femur

17 The thigh bone, or _____ , fits into the acetabulum of the hip to form the hip joint. The head of the femur is held tightly in the acetabulum by three very strong ligaments.

18 The femur has two projections, the *greater trochanter,* which is lateral, and the lesser trochanter which is medially placed.

greater trochanter

Both the lateral projection, the _____

_____ , and the medial projection,

lesser trochanter

the _____ , are important sites for muscle attachment.

19 Examine Fig. 5-2 and note that the medial condyle of the

68

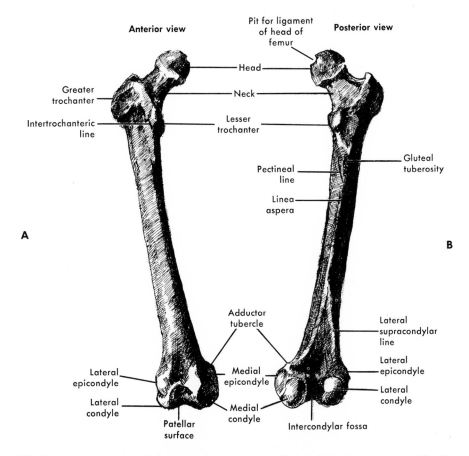

Anterior view

Pit for ligament
of head of
femur

Posterior view

Head

Greater
trochanter

Neck

Intertrochanteric
line

Lesser
trochanter

Gluteal
tuberosity

Pectineal
line

Linea
aspera

A

B

Adductor
tubercle

Lateral
supracondylar
line

Lateral
epicondyle

Medial
epicondyle

Lateral
epicondyle

Lateral
condyle

Medial
condyle

Lateral
condyle

Patellar
surface

Intercondylar fossa

FIG. 5-2. A, Anterior view of the femur. Note the angle of the shaft. B, Posterior view of the femur.

distal
distal

femur is on the (proximal/distal) end and the adductor tubercle is on the (proximal/distal) end.

posterior

20 The linea aspera of the femur is on the (anterior/posterior) surface.

greater and lesser
  trochanters

21 The intertrochanteric line runs between the (head and lesser trochanter/lesser trochanter and the medial epicondyle/ greater and lesser trochanters).

more

22 The neck of the femur is (more/less) pronounced than the neck of the humerus.

23 The musculature of the pelvic girdle performs the same functions for the hip joint as the muscles of the shoulder girdle perform for the shoulder joint. The range of motion is

more

(more/less) limited in the hip.

THE GLUTEI

24 The muscles of the "buttocks" are coarse fibered; and there are three of them:

gluteus maximus   gluteus medius   gluteus minimus

25 From their names, one could guess that, of the three gluteal

gluteus maximus

muscles, the _____ is the largest and the

gluteus minimus

_____ is the smallest.

26 The gluteus maximus (Fig. 5-3, *A*) originates on the

iliac

_____ bone of the pelvis and on the sacrum and
coccyx (a fragile continuation of the spinal column).

sacrum

27 From its origin on the ilium, the _____, and

coccyx

the _____, the gluteus maximus inserts onto the

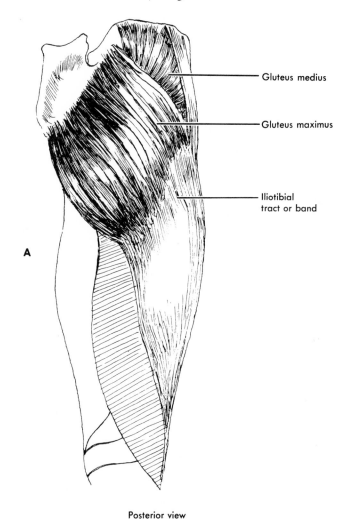

Gluteus medius

Gluteus maximus

Iliotibial
tract or band

A

Posterior view

FIG. 5-3. **A,** Gluteus maximus, posterior view. The muscle attaches both to the iliotibial band and to the gluteal tuberosity. **B,** The gluteus medius and minimus lie deep to the gluteus maximus. Note that the gluteus medius and minimus are shown in a lateral view and the maximus is shown in a posterior view.

70

femur

gluteal tuberosity

thigh, or _____ , bone at the gluteal tuberosity.

28 In addition to its insertion on the _____ , the gluteus maximus inserts into a large sheet of fascia on the lateral aspect of the thigh called the *iliotibial band* or tract.

29 The gluteus medius (Fig. 5-3, *B*) arises from the ilium

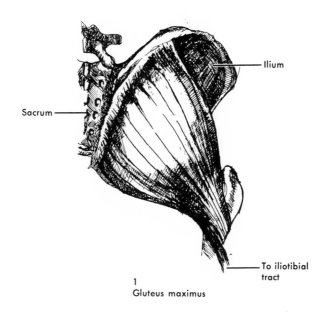

1
Gluteus maximus

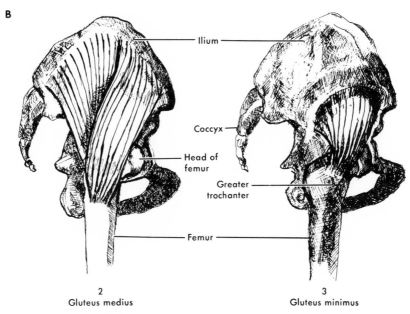

2
Gluteus medius

3
Gluteus minimus

FIG. 5-3, cont'd. For legend see opposite page.

71

somewhat inferior to the gluteus maximus. From its origin on

the _____ , the gluteus medius inserts onto the *greater trochanter.*

ilium

greater trochanter

30 The insertion of the gluteus medius on the _____

_____ is shared somewhat by the gluteus minimus (Fig. 5-3, *B*). The minimus inserts more anteriorly than the medius.

inferior

31 The gluteus minimus arises on the ilium (superior/inferior) to the gluteus medius.

32 Of the three gluteal muscles, the gluteus maximus is the most (superficial/deep/lateral).

superficial

33 The bone of origin held in common by the three gluteals is

ilium

the _____ .

34 The gluteus maximus functions to extend the thigh but is not used in walking on level ground where there is little resistance. Extension against resistance in which the gluteus maximus is involved is seen in (jogging/stair climbing/sit-ups/jumping).

stair climbing and jumping

35 The gluteus maximus also acts to extend the trunk when the legs are fixed, as in doing trunk raising exercises in the prone position. In the action of trunk raising from the prone

gluteus maximus

position, the _____ muscle reverses the usual roles of origin and insertion.

36 A second function of the gluteus maximus is lateral rotation of the thigh. Lateral rotation of the thigh is accomplished by turning the knee (inward/outward).

outward

37 The gluteus medius and minimus act in concert to abduct the thigh and rotate it medially. The abduction provided by the

medius; minimus

gluteus _____ and _____ is important in walking.

38 In walking the gluteus medius and minimus contract on the hip opposite to that of the leg being moved. They thus hold the hip line steady and prevent its sagging on the unsupported side by pulling the hip down to the femur on the (same/opposite) side (Fig. 5-4).

opposite

are

39 In the act of skating the gluteus medius and minimus (are/are not) important.

40 When a skating stride is taken with the left leg, the (right/left/both) gluteus medius and minimus are contracted.

left

41 The gluteus maximus is innervated by the *inferior gluteal* nerve. In the act of (walking/stair climbing) the gluteus

stair climbing

inferior gluteal

maximus is used, and it is stimulated by the _____

_____ nerve.

42 The gluteus medius and minimus are innervated by the

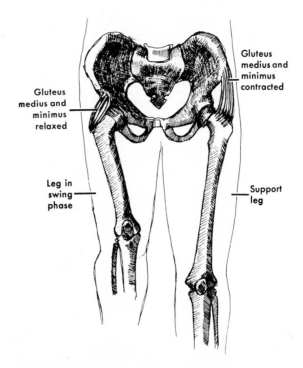

Gluteus
medius and
minimus
relaxed

Gluteus
medius and
minimus
contracted

Leg in
swing
phase

Support
leg

FIG. 5-4. The leg on the right in this illustration is supporting the body weight, and the gluteus medius and minimus on the same side are contracted to hold the pelvis level.

walking

superior gluteal nerve. In the act of (swimming the crawl/ walking) the gluteus medius and minimus are contracted by

superior gluteal

impulses from the _____ nerve.

LATERAL ROTATORS OF THE HIP

maximus

43 A set of small muscles of the hip provides lateral rotation to assist the gluteus (medius/maximus/minimus).

44 The lateral rotators (Fig. 5-5) are:

    piriformis
    gemellus superior
    gemellus inferior
    obturator internus
    quadratus femoris
    obturator externus

A mnemonic to help remember the lateral rotators of the hip is "pretty girls often quit outright." The plural "girls" indicates more than one muscle.

45 The lateral rotators serve in part to stabilize the hip joint. The group in the shoulder to which they have the greatest

rotator cuff muscles

similarity is the _____ .

46 The term *piriformis* means pear shaped. The piriformis

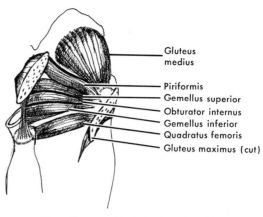

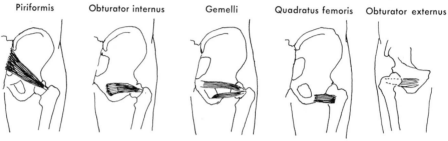

| Piriformis | Obturator internus | Gemelli | Quadratus femoris | Obturator externus |

FIG. 5-5. Lateral rotators of the femur. Note that the obturator externus runs behind the head and neck of the femur to reach the greater trochanter.

deep

    originates on the sacrum and ilium (superficial/deep) to the gluteus maximus.

sacrum

47 The piriformis from its origin on the _____ and

ilium

    _____ inserts onto the posterior surface of the greater trochanter.

    48 The gemelli (= twins) arise from the ischial spine and ischial tuberosity, respectively, and lie above and below the ob-

gemelli

    turator internus muscle. The twins, or_____, insert onto the tendon of the muscle between them, the

obturator internus

    _____ muscle.

    49 The obturator internus is obviously associated with the

obturator

    _____ foramen of the pelvis.

    50 The obturator foramen is closed by its membrane, and the membrane is the origin of the obturator internus. The obturator internus tendon picks up the gemelli and then

greater trochanter

    inserts near the piriformis on the _____

    _____ of the femur.

    51 The obturator internus inserts on the medial side of the greater trochanter, whereas the intertrochanteric crest is the

74

quadratus femoris	insertion of the next muscle of the group, the _____ .
	52 The quadratus femoris arises on the ischial tuberosity of the
intertrochanteric crest	ischium and then inserts on the _____ _____ of the femur.
	53 The obturator externus occupies for its origin the opposite surface to that supplying origin to the obturator internus, the
obturator	_____ membrane.
	54 The obturator externus tendon runs backward behind and inferior to the neck of the femur to insert into the trochanteric fossa. The circuitous route behind the neck of
greater trochanter	the femur to its insertion into the _____ _____ enables the obturator externus to contribute to the major function of this group, that is,
lateral	(medial/lateral) rotation.
quadratus femoris	**55** The most inferior of the lateral rotators is the _____ _____ .
	56 The only muscle of the group that arises from the front of the
obturator externus	pelvic girdle is the _____ .
	57 In this respect the obturator externus is similar to one of the
subscapularis	rotator cuff muscles of the shoulder, the _____ .
	58 The term *piriformis* means pear shaped, *obturator* means a stopper, or plug, and *gemellus* means twin. A glance at the muscles showing their shape and position indicates that the names are fitting.
quadratus femoris	**59** Note that two muscles, the _____
gemellus inferior	and the _____ , originate on the ischial tuberosity.
piriformis	**60** The pear-shaped muscle, the _____ , is the only muscle of this group to originate on the sacrum.
	61 The innervation of the rotator muscles of the hip is largely by sacral and lumbar nerves, the fourth and fifth lumbar nerves to the sixth sacral nerve. The obturator externus is stimulated by the posterior branch of the obturator nerve.
lateral rotators and gluteus maximus	**62** The action of turning the knee outward necessitates the contraction of the (lateral rotators/gluteus medius and minimus/gluteus maximus).
	63 The nerves acting to contract these muscles are the
inferior gluteal	_____ nerve for the gluteus maximus and
L4; S2	the lumbar-sacral nerves from _____ to _____ , which stimulate the lateral rotators, plus the

obturator	_____ nerve of the obturator externus.
	64 The quadratus femoris and obturator externus provide an additional function. From their position, one might deduce
adduct	that they (abduct/adduct/flex) the thigh.
	65 The obturator internus acts in abducting the thigh along with
piriformis	the (gemelli/piriformis).
	66 The blood supply to the area is largely via the *superior* and *inferior gluteal arteries* and the *obturator* artery.
	67 Thus the nomenclature of the arteries in the area parallels
inferior gluteal	that of the nerves to the muscles: the _____
inferior gluteal	nerve and the _____ artery to the gluteus
superior gluteal	maximus and the _____ nerve and
superior gluteal	_____ artery to the gluteus medius and minimus.

HIP FLEXORS

68 The thigh is flexed by four strong muscles, the IRST muscles:

> iliopsoas (made up of two muscles, the iliacus and psoas)
> rectus femoris (the "kicking muscle")
> sartorius (the "tailor's muscle" because tailors used to sit cross-legged)
> tensor fasciae latae

	69 Examine Figs. 5-6 and 8-4 and note that the psoas muscle arises from the lumbar vertebrae and inserts onto the lesser tro-
lumbar vertebrae	chanter. As the psoas proceeds from its origin on the _____ _____ , it is joined by the iliacus.
	70 The iliacus arises from the ilium as the name suggests and
lesser trochanter	inserts in common with the psoas onto the _____ _____ .
	71 The common tendon of the two muscles passes over the brim
pubis	of the pelvis and lies atop the (ilium/ischium/pubis).
rectus femoris	**72** The "kicking muscle," or _____ (Fig. 5-7), lies on the anterior aspect of the thigh and arises from two sources: the anterior inferior iliac spine and the superior margin of the acetabulum.
anterior inferior iliac	**73** From its origins on the pelvis at the _____ _____ spine and the superior
acetabulum	margin of the _____ , the rectus femoris inserts on the *tibial tuberosity.*

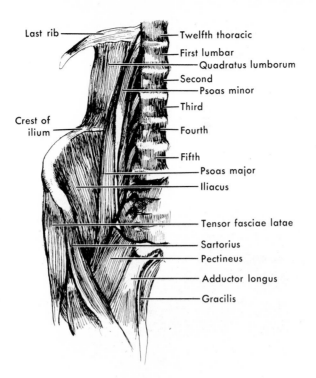

Last rib
Twelfth thoracic
First lumbar
Quadratus lumborum
Second
Psoas minor
Third
Crest of
ilium
Fourth
Fifth
Psoas major
Iliacus
Tensor fasciae latae
Sartorius
Pectineus
Adductor longus
Gracilis

FIG. 5-6. The psoas and iliacus have a common tendon of insertion to the lesser trochanter. The pectineus arises from the superior ramus of the pubis, and the adductor magnus and gracilis arise from the inferior ramus.

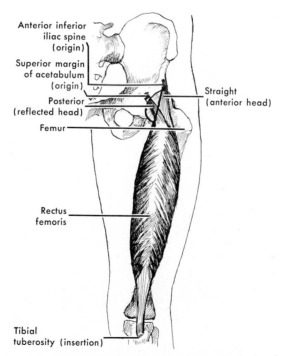

Anterior inferior
iliac spine
(origin)
Superior margin
of acetabulum
(origin)
Straight
(anterior head)
Posterior
(reflected head)
Femur
Rectus
femoris
Tibial
tuberosity (insertion)

FIG. 5-7. The rectus femoris crosses two joints. The reflected, or posterior, head arises from the rim of the acetabulum.

77

74 The point of insertion of the rectus femoris on the

tibial tuberosity

_____ is on the tibia, or shinbone just below the kneecap.

tailor's

75 The sartorius, nicknamed the "_____ muscle," originates on the anterior superior iliac spine and

superior

therefore is (superior/inferior) to the origin of the rectus
femoris.

anterior superior iliac

76 From its origin on the _____

_____ spine, the sartorius stretches down to insert
onto the tibia medial to the tibia tuberosity.

tibia

77 The insertion of the sartorius on the _____ in-

two

dicates that the muscle crosses (one/two/three) joints.

all three actions (in the
   push-ups the iliopsoas
   prevents the thighs
   from sagging)

78 The iliopsoas is the strongest of the thigh flexors and is
innervated by the second and third lumbar nerves. In the
action of (hurdling/sit-ups/push-ups) the iliopsoas functions,

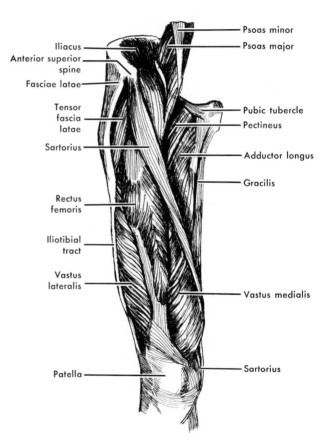

Iliacus
Anterior superior spine
Fasciae latae
Tensor fascia latae
Sartorius
Rectus femoris
Iliotibial tract
Vastus lateralis
Patella

Psoas minor
Psoas major
Pubic tubercle
Pectineus
Adductor longus
Gracilis
Vastus medialis
Sartorius

FIG. 5-8. Quadriceps, anterior view. Note the sartorius running diagonally
across the three visible muscles of the quadriceps: the vastus lateralis, rectus
femoris, and vastus medialis.

second and third	stimulated by the _____ lumbar nerves.
	79 The rectus femoris, which is innervated by the *femoral* nerve acts not only to flex the hip but also to extend the
kicking	knee—hence its nickname, the " _____ muscle."
femoral	**80** The _____ nerve of the rectus femoris also stimulates the sartorius, which is the longest muscle in the body.
anterior superior iliac	**81** The sartorius from its origin on the _____
	_____ spine can act to flex the hip and draw the knee up and laterally—hence its nickname, the "tailor's muscle."
four	**82** The tensor fasciae latae, when stimulated by the inferior gluteal nerve, acts on the lateral fascia (fasciae latae) of the thigh to provide hip flexion. It is thus one of (one/two/three/ four) muscles studied thus far that flex the hip.
medial	**83** Judging from its position (Fig. 5-8), one could say that the tensor fasciae latae also provides (medial/lateral) rotation of the hip.

79

$\mathbb{6}$ The thigh

The thigh is divided into three compartments: anterior, posterior, and medial.

ANTERIOR COMPARTMENT

quadriceps

1 The muscles of the front of the thigh, or anterior compartment, are four in number and thus named the (triceps/biceps/quadriceps).

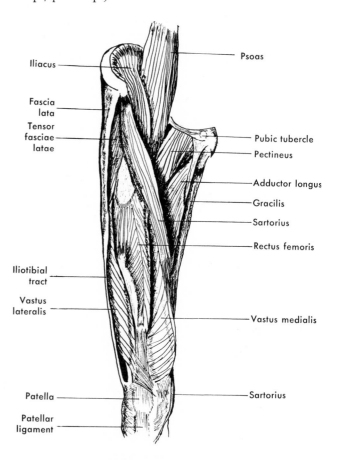

FIG. 6-1. Front of the thigh, showing most of the muscles of the anterior compartment, the quadriceps. The sartorius is the longest muscle in the body.

2 All the quadriceps are innervated by the *femoral* nerve. In Fig. 6-1 three of the four quadriceps are shown. They are the

flexor

rectus femoris, a (flexor/extensor) of the hip joint and an

extensor

(flexor/extensor) of the knee joint, the *vastus lateralis* and the *vastus medialis.*

3 The *vastus intermedius,* a third member of the vasti group, lies deep to the rectus femoris between the other vasti muscles.

4 The quadriceps have a common tendon of insertion, the *patellar* ligament, which inserts with the rectus femoris onto

tibial tuberosity

the _____ .

5 The patella, or kneecap, is a *sesamoid* bone that is encased completely by the quadriceps tendon. The kneecap, or

patella

_____ , is situated between the epicondyles of the tibia.

6 The vastus lateralis originates on the upper portion of the lateral aspect of the femur, including the greater trochanter

patellar

and the gluteal tuberosity, and inserts via the _____ ligament onto the tibial tuberosity.

between

7 The vastus intermedius lying (between/above/below) the other vasti muscles originates on the upper half of the front of the femur.

8 The vastus medialis arises on the medial aspect of the upper femur on the *linea aspera* and *intertrochanteric line.* Like the other quadriceps, the vastus medialis is innervated by the

femoral

_____ nerve.

intertrochanteric line

9 From its origin on the linea aspera and the _____

_____ , the vastus medialis runs down to join the common tendon of the quadriceps; however, the distal fibers of the vastus medialis run almost at right angles to the long axis of the femur.

10 The one muscle of the quadriceps whose fibers do not follow

vastus medialis

the long axis of the femur is the _____ . The medial pull on the patella by the vastus medialis helps off-set a tendency for the kneecap to dislocate laterally.

extension

11 The quadriceps are chiefly responsible for (flexion/extension/rotation) of the knee.

all three

12 They are used in (walking/running/skating).

13 The extension of the knee joint is accomplished by the

quadriceps

_____ group of muscles, innervated by the

femoral

_____ nerve.

14 The muscles of the back of the thigh work in opposition to the quadriceps; therefore they perform the action of

flexion

_____ on the knee.

rectus femoris

15 One muscle of the quadriceps, the _____

_____ , flexes the hip. The muscles of the back of

extend

the thigh do the opposite. They _____ the hip.

POSTERIOR COMPARTMENT

flexion; extension

16 The muscles of the back of the thigh are known collectively as the *hamstrings* (Fig. 6-2). The hamstrings provide two functions: (flexion/extension) of the knee and (flexion/extension) of the hip.

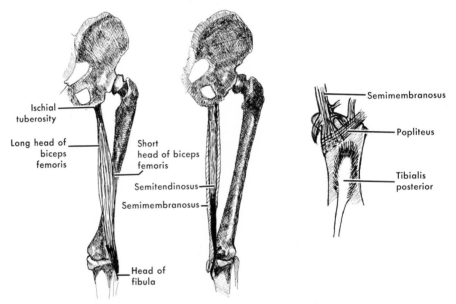

FIG. 6-2. Muscles of the posterior compartment of the thigh. The widespread insertion of the semimembranosus enlarged on the right. Note details of the insertion of the semimembranosus.

posterior

17 The hamstrings situated on the (posterior/lateral/anterior) surface of the thigh are the:

    biceps femoris
    semitendinosus
    semimembranosus

two

18 With one exception, the hamstrings all originate from the ischial tuberosity. The exception is the biceps because it has (three/two/one) heads of origin.

ischial

19 The long head of the biceps originates on the _____ tuberosity, and the short head originates on the linea aspera.

20 By common tendon the biceps femoris runs in a lateral oblique direction to insert onto the head of the *fibula* (a smaller companion bone to the tibia) to help provide the

flexion

action of _____ of the knee.

82

fibula; lateral	**21** The _____ where the biceps inserts is (lateral/ medial) to the tibia.
	22 The semitendinosus inserts onto the medial aspect of the tibia
ischial tuberosity	from its origin on the _____ .
	23 The semitendinosus, as its name implies, is made up for half
tendon	its length of _____ .
	24 Lying deep to the semitendinosus is the last member of the
semimembranosus	hamstrings, the _____ .
	25 The semimembranosus, like the semitendinosus, inserts onto
tibia	the _____ ; however, its insertion is much more diverse than that of the semitendinosus.
	26 Some portions of the semimembranosus tendon insert on the anterior surface of the tibia, whereas other portions are posterior. Together with the rest of the hamstrings, the semimembranosus provides flexion of the knee and
extension	_____ of the hip.
	27 All the hamstrings are innervated by a part of the *sciatic* nerve called the *tibial* nerve, with one exception. The exception is
short head of the biceps	the _____ muscle, which is the only hamstring not originating on the ischial tuberosity.

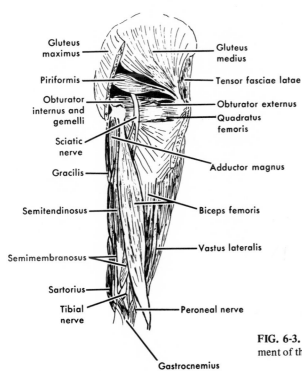

Gluteus maximus
Gluteus medius
Piriformis
Tensor fasciae latae
Obturator internus and gemelli
Obturator externus
Quadratus femoris
Sciatic nerve
Gracilis
Adductor magnus
Semitendinosus
Biceps femoris
Vastus lateralis
Semimembranosus
Sartorius
Tibial nerve
Peroneal nerve
Gastrocnemius

FIG. 6-3. Muscles of the posterior compartment of the thigh, glutei, and lateral rotators.

28 The short head of the biceps is innervated by the *peroneal* nerve, which, like the tibial nerve, is part of the larger

sciatic

_____ nerve.

29 In the action of running, the hamstrings are used on the

backward

(forward/backward) thrust of the thigh and knee. (See Fig. 6-3.)

30 In the backward thrusting action the hamstrings are innervated by the two portions of the sciatic nerve, the

tibial; peroneal

_____ nerve and the _____ nerve.

31 The action of lifting the heel to the buttocks is brought about

hamstrings

by contraction of the _____ , innervated by

sciatic

the _____ nerve.

MEDIAL COMPARTMENT

medial

32 The adductor muscles are those of the (medial/lateral) compartment of the thigh (Fig. 6-4).

33 The adductor muscles of the medial compartment are from medial to lateral:

gracilis	adductor brevis
adductor magnus	pectineus
adductor longus	obturator externus

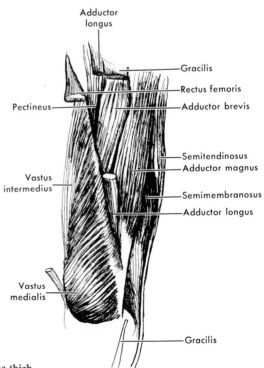

FIG. 6-4. Medial view of the adductors of the thigh.

84

They might be termed the GALBPO muscles, and all but the pectineus are innervated by the *obturator nerve*.

34 The gracilis was once termed the guardian of virginity. It is a weak muscle originating on the inferior ramus of the pubis

obturator and is innervated by the _____ nerve.

inferior ramus 35 From its origin on the _____ of the pubis, the gracilis inserts onto the medial aspect of the upper tibia.

36 The gracilis, to reach its insertion on the medial aspect of the

upper tibia _____, must cross two joints.

37 The adductor magnus originates on the ischial tuberosity and the ischiopubic ramus. It has two somewhat distinct parts—an adductor and an extensor portion. The extensor portion arises

ischial from the _____ tuberosity, and the adductor portion arises from the ischiopubic ramus.

38 The extensor portion of the adductor magnus inserts onto the *adductor tubercle* and linea aspera of the posterior surface of

hamstrings the femur; thus its fibers really run with the (hamstrings/ quadriceps) and like them are innervated by the sciatic nerve.

39 The adductor portion has fibers running almost horizontally

ischiopubic from its insertion on the _____ ramus to insert into the posterior surface of the femur near the gluteal tuberosity.

40 The adductor longus arises on the superior ramus of the pubis and inserts with part of the adductor magnus onto the

linea _____ aspera.

laterally 41 The adductor fibers thus run down and (medially/laterally)

pubis from their origin on the _____.

brevis 42 The third adductor, the adductor _____ , arises on the inferior ramus of the pubis and inserts into

linea aspera the _____ with the adductor longus. The adductor brevis lies deep to the adductor longus and pectineus.

43 The pectineus (= comb) arises from the pectineal line of the superior ramus of the pubis and inserts onto the pectineal line of the femur. The pectineus thus runs from one

pectineal; pectineal _____ line to a second _____ line and descends behind the lesser trochanter to get to its insertion.

44 The obturator internus was previously discussed as a mem-

lateral ber of the (medial/lateral) rotators of the thigh. It

obturator originates on the _____ membrane and

greater inserts onto the (greater/lesser) trochanter.

45 The pectineus, like the obturator externus, is also a member of another group. It could be classified with the quadriceps

femoral

because like them it is innervated by the _____

rectus femoris

nerve and like the _____ is a hip flexor.

46 Clinging to a horse with the knees involves the muscles of the

medial, or adductor

_____ , compartment, innervated

obturator

principally by the _____ nerve.

47 The adductor muscles are:

gracilis

_____

adductor magnus

_____

adductor longus

_____

adductor brevis

_____

pectineus

_____

obturator externus

_____

BLOOD SUPPLY TO THE THIGH

48 The blood supply to the entire thigh (Fig. 6-5) is from the *femoral* artery and its branches.

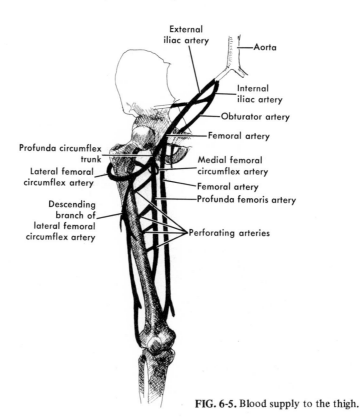

FIG. 6-5. Blood supply to the thigh.

49 The femoral artery reaches the rear thigh by piercing the muscle attached to the adductor tubercle, the adductor

magnus
_____ , through the *adductor canal.*

50 The femoral artery also pierces the adductor magnus with smaller perforating branches to reach the posterior thigh, as

adductor
well as having its main branch go through the _____ canal.

51 Blood to the medial compartment is supplied by the

femoral
_____ artery.

52 On Fig. 6-6 note the three compartments of the thigh, with the muscles indicated schematically at about the middle of the thigh. Name them, keeping in mind size and position as clues to their identity. There is one nerve indicated.

1—Vastus lateralis	**5**—Biceps femoris	**9**—Adductor magnus
2—Vastus intermedius	**6**—Semitendinosus	**10**—Gracilis
3—Vastus medialis	**7**—Semimembranosus	**11**—Adductor longus
4—Rectus femoris	**8**—Sciatic nerve	**12**—Sartorius

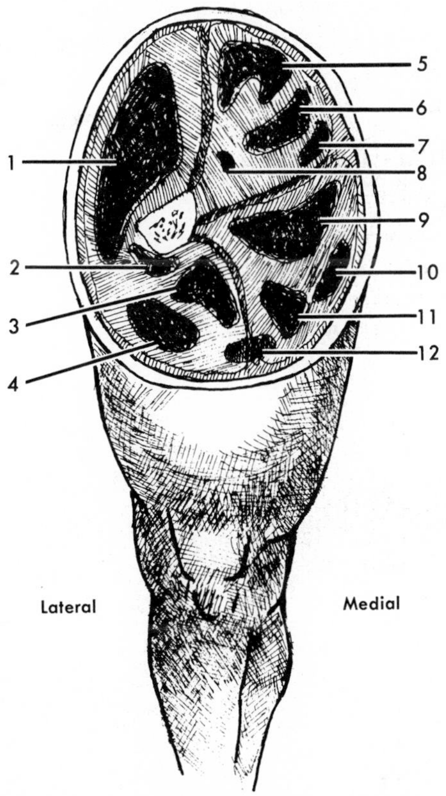

FIG. 6-6. Horizontal section of the midthigh, showing the three compartments, their muscles, and one major nerve in the posterior compartment.

THE KNEE

hinge

laterally

capsule

synovial

menisci

ligaments

53 An examination of the movement of your knee reveals that it is a (ball-and-socket/hinge) joint.

54 The tibia is straight, but the femur slants upward from the knee (laterally/medially) at an angle of 10 to 12 degrees.

55 The knee is a synovial joint and therefore must possess a (ligaments/capsule). The latter starts on the femur above the intercondylar fossa and attaches on the tibia at the condyles.

56 The capsule is lined with synovium that secretes

_____ fluid.

57 Within the knee joint resting on the tibia are two *menisci*, or cartilaginous discs—a medial and a lateral. The discs, or

_____ , of the knee are frequently damaged in such sports as football and hockey.

58 The capsule of the knee is strengthened as in other joints by

_____ .

59 The ligaments of the knee joint (Fig. 6-7) are:

tibial (medial) collateral ligament
fibular (lateral) collateral ligament
oblique popliteal ligament (not shown)
anterior cruciate ligament
posterior cruciate ligament

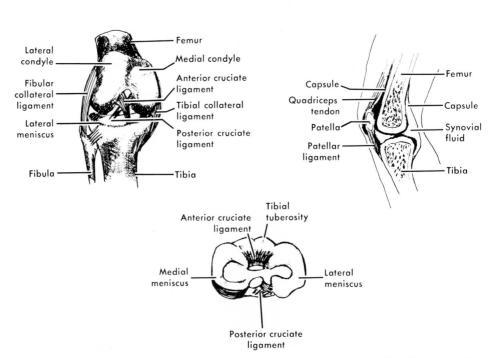

FIG. 6-7. The knee joint is supported by several ligaments, a capsule, and several tendons. Note the position of the menisci.

88

A mnemonic to remember them is "two fellows obliged Aunt Polly."

60 Examine Fig. 6-7 and note that the tibial collateral ligament is on the (anterior/medial/lateral) aspect of the knee and runs

medial

from the condyle of the _____ to the condyle of

femur

the _____ .

tibia

61 The fibular collateral ligament attaches to the head of

fibula; femur

the _____ and the condyle of the _____ .

side-to-side

62 The tibial and fibular collateral ligaments prevent (front-to-back/side-to-side/rotary) movements in the knee when the knee is extended.

63 The oblique popliteal ligament runs from the medial posterior aspect of the tibia upward to the lateral condyle of the femur.

posterior

It is on the (anterior/posterior/lateral) aspect and is a reflection of the tendon of the semimembranosus.

64 From its position one could deduce that the purpose of the

overextension

oblique popliteal ligament is to prevent (overflexion/overextension/overrotation).

65 The cruciate (cruciate = cross) ligaments run from the

tibia

articular surface of the (tibia/fibula) upward to the femur.

66 The anterior cruciate ligament arises anterior to the intercondyle ridge and attaches to the internal surface of the lateral femoral condyle. Its origin lies between the two

menisci

_____ lying atop the tibia.

67 The posterior cruciate ligament arises posterior to the intercondylar ridge of the tibia and crosses the other cruciate ligament to attach to the internal surface of the medial femoral condyle. The anterior cruciate ligament runs to the

lateral

_____ condyle of the femur, whereas the

medial

posterior runs to the _____ condyle.

68 The most likely function of the cruciates would appear to be

anteroposterior displacement

the prevention of (anteroposterior displacement/side-to-side displacement/movement of the menisci) during flexion and extension.

69 The knee joint is not a stable joint, and its ligaments are reinforced by the muscles of the thigh and the leg. The ligaments wrinkle somewhat in flexion. The knee joint is thus

extension

most stable in (extension/flexion).

89

7 The leg and foot

1 The two bones of the leg, that is, that part of the lower limb below the knee that is correctly called the leg, are the _____ and the _____.

tibia

fibula

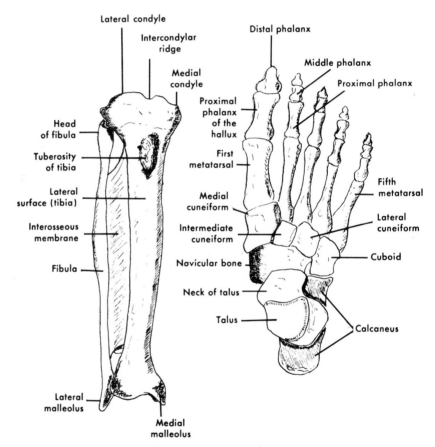

FIG. 7-1. Bones of the right leg and foot (top view). Note that the foot is made larger and out of proportion to the leg bones to show details more clearly.

2 Examine Fig. 7-1 and note that they are joined by the _____ membrane.

interosseous

distal	**3** The malleoli are found on the (proximal/distal) aspect.
talus	**4** The tibia and fibula fit over the (calcaneus/cuboid/talus) (Fig. 7-9).
calcaneus	**5** The largest bone of the foot is the _____.
	6 The talus sits atop the calcaneus and articulates with the
navicular	_____ bone anteriorly.
three	**7** Note that there are (two/three/four) cuneiform and one cuboid bone.
	8 The talus, calcaneus, cuboid, three cuneiform, and navicular bones make up the tarsal bones, and they are analogous to
carpal	the _____ bones of the wrist.
	9 The next distal to the carpal bones of the wrist are the metacarpals. In the foot the corresponding bones are the
metatarsals	_____ .
	10 There are eight carpal bones in the wrist but only
seven	_____ tarsal bones in the foot. The
do	metacarpals and metatarsals (do, do not) have corresponding numbers.
	11 The phalanges of the foot are parallel to those of the hand except that the middle phalanx of the foot is considered to
two	be number (two/four/three), whereas in the hand, the
three	middle phalanx is number (two/three/four).
	12 The thumb, or pollux, of the hand has a counterpart in the
hallux	foot in the form of the big toe, or _____.

MUSCLES OF THE LEG

13 Exactly as the muscles of the thigh are divided into a number of compartments, or groups, so the muscles of the

three   leg are divided into (two/three/four) groups (Fig. 7-2).

14 The basis of their division, as in the thigh, is that muscles

have the same   within a group (serve the same function/have the same blood
innervation   supply/ have the same innervation).

Anterior compartment

15 The anterior compartment of the leg contains from tibia laterally:

    tibialis anterior
    extensor digitorum longus
    peroneus tertius
    extensor hallucis longus

The mnemonic **TEPE** (teepee) will help one remember them

laterally   in order from the tibia (laterally/medially).
anterior   **16** The tibialis anterior originates on the tibia on the (anterior/ posterior) surface and from the interosseous membrane.

17 From its origin on the anterior surface of the tibia and the

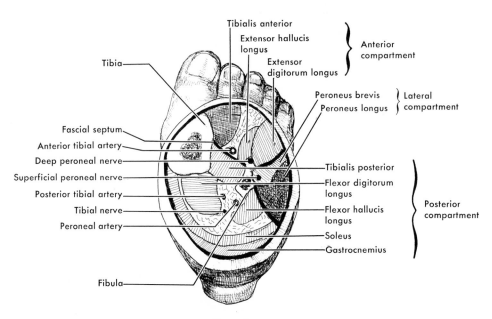

FIG. 7-2. Compartments of the leg. The fascial septa divide the leg into three compartments. This section is taken about midcalf and will not show the peroneus tertius.

interosseous membrane

_____ , the tibialis anterior inserts onto the medial cuneiform and the first metatarsal. Both insertions are on the medial surfaces.

18 The second muscle of the anterior compartment, the

extensor digitorum longus

_____ , originates on the upper three fourths of the fibula and inserts into the membranous expansion of the lateral four toes.

19 To prevent the tendons of the extensor digitorum longus from "bowing" as they run from their origin on the

fibula; lateral four toes

_____ to their insertion on the _____

_____ , a two-part extensor retinaculum holds them at the ankle (Fig. 7-3).

20 The peroneus tertius originates from the lower portion of the fibula and adjacent interosseous membrane as part of the

extensor digitorum longus

_____ muscle.

21 The peroneus tertius runs from its origin on the

fibula

_____ to the base of the fifth metatarsal.

22 The extensor hallucis longus arises from a narrow area on the middle of the fibula and from the interosseous mem-

big
medial

brane and runs to the distal phalanx of the _____ toe.

23 Thus the extensor hallucis longus runs to the (medial/lateral) side of the foot, and the peroneus tertius runs to the

lateral

(medial/lateral) side.

24 The point of origin common to all the muscles of the

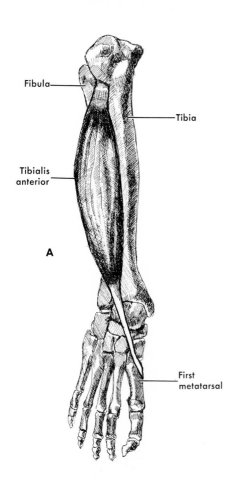

A

Fibula

Tibia

Tibialis
anterior

First
metatarsal

FIG. 7-3. A, Tibialis anterior. **B,** Extensor
digitorum longus. The enlargement shows the
tendons passing under the two portions of the
flexor retinaculum.

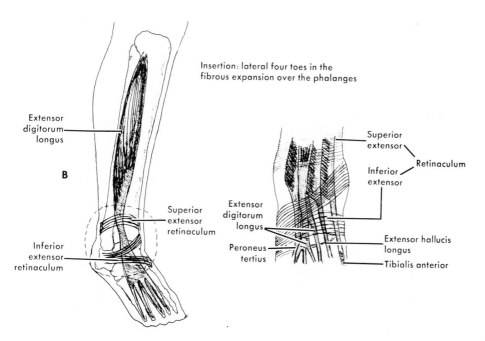

Insertion: lateral four toes in the
fibrous expansion over the phalanges

Extensor
digitorum
longus

B

Inferior
extensor
retinaculum

Superior
extensor
retinaculum

Superior
extensor

Inferior
extensor

Retinaculum

Extensor
digitorum
longus

Peroneus
tertius

Extensor hallucis
longus

Tibialis anterior

interosseous membrane

dorsiflex
invert

lateral

everting

dorsiflexion

everting

dorsiflexing

dorsiflexion; inversion

dorsiflexion

deep peroneal

fibula

base of the first meta-
    tarsal and medial
    cuneiform

lateral
lower

superficial peroneal
evert

extensor compartment is the _____ .

25 From an examination of Fig. 7-3 one could deduce that the tibialis anterior acts to (dorsiflex/plantar flex) the foot and also to (invert/evert) the foot.

26 Palpate your own tibialis anterior (right leg), which lies just (lateral/medial) to the sharp edge of your "shinbone," and perform the act of dorsiflexion and inversion.

27 The extensor digitorum longus may oppose the action of the tibialis anterior by (everting/plantar flexing) the foot and

assist it in the action of _____.

28 The peroneus tertius assists the extensor digitorum longus

with which it is closely associated in the act of _____

and _____ the foot.

29 The extensor hallucis longus assists the tibialis anterior in its

actions of _____ and _____
of the foot, and, of course, the extensor hallucis acts on the

big toe to produce _____ .

30 The muscles of the anterior compartment are innervated by the *deep peroneal* nerve and are important in walking and running. During the act of walking or running, the

_____ nerve stimulates the muscles of the anterior compartment to help clear the toes from the ground as the foot is brought forward after taking a step.

Lateral compartment

31 The lateral compartment of the leg contains the peroneus (= fibula) longus and peroneus brevis muscles, and both are attached to the (tibia/fibula).

32 The insertions of the muscles of the lateral compartment are on opposite sides of the foot (Fig. 7-4). The peroneus longus originates on the upper part of the fibula and inserts on the base of the first metatarsal and medial cuneiform. To reach

its insertion on the _____
the peroneus longus passes under the foot.

33 The peroneus brevis originates from the lower portion of the fibula and inserts on the (lateral/medial) aspect of the foot.

34 The peroneus brevis from its origin on the (upper/lower) fibula inserts onto the base of the fifth metatarsal.

35 The peronei are supplied by the *superficial peroneal* nerve; from an examination of their insertions, it can be seen that,

when stimulated by the _____
nerve, they act to (invert/evert) the foot.

36 The peronei, since their insertions are on the underside, or

94

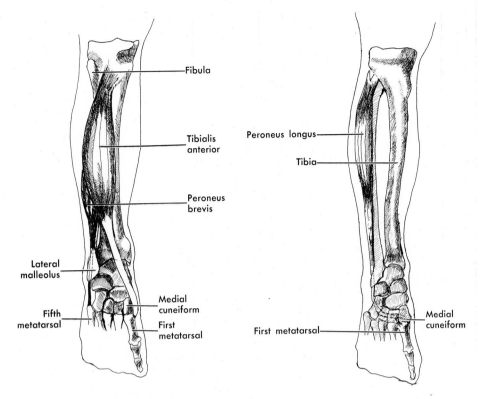

FIG. 7-4. Peronei muscles in relation to the tibialis anterior. The insertion of the peroneus longus is on the opposite aspect of the first metatarsal from the insertion of the tibialis anterior.

plantar flexion

plantar surface, of the foot, also contribute to the action of (plantar flexion/dorsiflexion).

37 In the action of walking or running, the peronei are

plantar flexion

important for the actions of _____

eversion

and _____ and are stimulated by the

superficial peroneal

_____ nerve.

Posterior compartment

38 The posterior compartment contains, from superficial to deep, the following muscles:

> gastrocnemius
> soleus
> flexor digitorum longus
> flexor hallucis longus
> tibialis posterior

The mnemonic for these muscles is "going slowly fully flexes toes."

39 The gastrocnemius originates on the femur above the

most superficial

condyles

tarsal

Achilles

calcaneus; gastrocnemius

flexor digitorum
longus

condyles and is the (most superficial/deepest) of the muscles of the posterior compartment.

40 From its insertion above the _____ of the femur, the gastrocnemius inserts via the *Achilles* tendon onto the back of one of the (metatarsal/tarsal) bones, the *calcaneus.*

41 The _____ tendon of the gastrocnemius is held in common by the soleus.

42 The soleus originates from the upper portion of the tibia and fibula and then inserts via the Achilles tendon onto the

_____ with the _____ muscle.

43 The gastrocnemius and soleus together make up the *triceps surae* (= calf) (Fig. 7-5) and are superficial to the next

muscle of the compartment, the _____

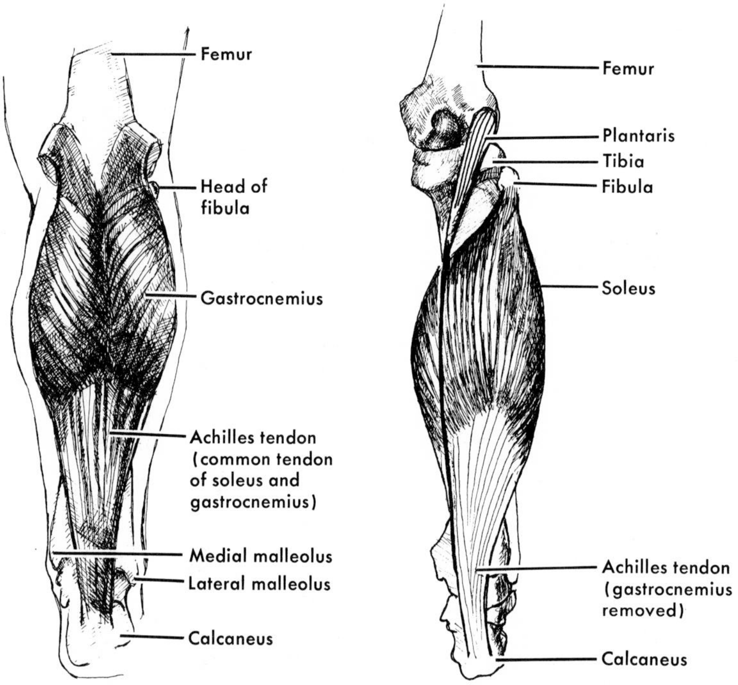

_____ .

FIG. 7-5. Triceps surae. The plantaris is not considered one of the triceps surae. The name triceps arises from the two heads of origin of the gastrocnemius and the deeper lying soleus. The muscles share a common tendon, the Achilles tendon.

44 The flexor digitorum longus originates on the tibia below

tibia

the soleus. From its origin on the _____ , the flexor digitorum longus inserts onto the distal phalanges of the lateral four toes.

45 The fact that the flexor digitorum longus inserts onto the

lateral four toes   distal phalanges of the _____

extensor digitorum   makes it the antagonist of the _____
longus
_____ of the anterior compartment.

46 The flexor hallucis longus originates on the lower portion of

posterior   the fibula on the (anterior/posterior) surface.

47 The flexor hallucis longus runs from its origin on the lower

fibula   _____ obliquely across the sole of the foot to
the base of the distal phalanx of the big toe.

tibialis posterior   **48** The last muscle of the group, the _____,
originates on both bones of the leg and the interosseous
membrane.

49 The tibialis posterior runs behind the medial malleolus from

tibia; fibula   its origin on the _____ , _____ ,

interosseous membrane   and _____ to a very extensive
insertion on the sole of the foot.

50 The insertion of the tibialis posterior covers the cuboid,

tarsals   cuneiform, and navicular bones, which are (tarsals/metatar-
sals).

51 The insertion of the tibialis posterior also goes to the lateral
four metatarsals.

two   **52** The gastrocnemius crosses (three/one/two) joints, and the
does not   soleus (does also/does not).

flex   **53** The gastrocnemius must act to (flex/extend) the knee joint
plantar flex   and (plantar flex/dorsiflex) the ankle.

54 Palpate your own Achilles tendon and note its prominence.
In older people it is not uncommon for unaccustomed
violent exercise to snap the tendon. The position of the

plantar flex   tendon indicated that the soleus acts to _____

_____ the ankle.

55 The triceps surae are innervated as are all the muscles of this
compartment by the *tibial* nerve. In walking or running, the

plantar flexion   triceps surae are important for their action of _____

tibial   _____ and are stimulated by the _____
nerve.

56 The flexor digitorum longus, judging from its insertion, is

plantar flexion   responsible for _____ of the toes.

57 Plantar flexion of the hallux is also the task of the flexor

tibial   hallucis longus when it is stimulated by the _____
nerve.

58 The flexor digitorum longus and flexor hallucis longus are

push-off

important in walking or running during the (heel strike/ push-off) phase.

59 The tendon of the tibialis posterior passes behind the (medial/lateral) malleolus. Therefore the muscle acts as an (invertor/evertor) of the foot, as well as a plantar flexor.

medial
invertor

60 Note on Fig. 7-6 that there are five tendons passing behind the malleoli, (four/two/three) behind the lateral, and (three/ two/one) behind the medial malleolus.

two; three

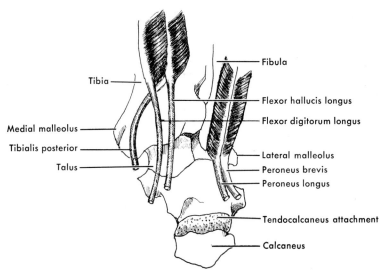

FIG. 7-6. Right ankle, posterior view. Some of the muscles of the posterior and lateral compartments pass behind the malleoli. Note that the deeper muscles of the posterior compartment pass medially to reach the sole of the foot.

61 Two other muscles associated with this area need mention. They are the popliteus and plantaris.

62 The plantaris is a small muscle with an extremely long tendon. It may be absent altogether and is unimportant. It is

palmaris longus

analogous to the _____ muscle of the flexor group of the forearm.

63 The popliteus arises from a pit on the lateral aspect of the lateral condyle of the femur and inserts into two areas of the (anterior/posterior) knee (Fig. 6-2).

posterior
lateral

64 From its origin on the (medial/lateral) surface of the femur, the popliteus runs obliquely down and medial to insert onto the lateral meniscus and then to the tibia next to the soleus.

65 The popliteus rotates the femur laterally when the leg is fixed and rotates the tibia medially when the femur is fixed. It is innervated by the nerve of the posterior compartment,

tibial

the _____ nerve.

66 The action of the popliteus is necessary because in full

extension of the knee, one or the other bone rotates (depending on which bone is fixed) in the last few degrees of straightening, and the popliteus acts to "unscrew" the joint slightly before flexion takes place.

BLOOD SUPPLY TO THE LEG

67 Examine Fig. 7-7 and note that the blood supply to the

anterior tibial

anterior compartment is from the _____ artery.

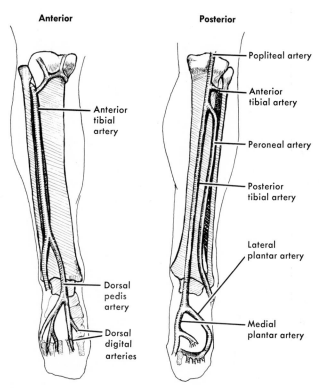

Anterior

Anterior tibial artery

Dorsal pedis artery

Dorsal digital arteries

Posterior

Popliteal artery

Anterior tibial artery

Peroneal artery

Posterior tibial artery

Lateral plantar artery

Medial plantar artery

FIG. 7-7. Blood supply to the leg and foot. The anterior tibial artery passes through a hiatus in the interosseous membrane to reach the anterior surface.

68 The blood supply to the posterior compartment is from the

posterior tibial

_____ artery.

69 The blood supply to the lateral compartment is from the

peroneal

_____ artery.

THE FOOT AND ANKLE

70 The ligaments supporting the ankle joint are a frequent site of injury in sports, especially those of the lateral aspect of the ankle.

71 The ligaments of the ankle most often injured are the (lateral/medial) ligaments.

72 The movement that would damage the lateral ligaments is (inversion-supination/eversion-suprination/inversion).

73 Examine Fig. 7-8 and note that the (medial/lateral/anterior) ligament of the ankle is the largest.

lateral

inversion-supination
medial

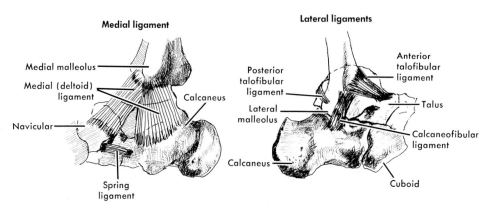

FIG. 7-8. Ligaments of the ankle.

74 The spring ligament supports part of the talus from the (superior/inferior) aspect.

75 The spring ligament is on the inferior medial aspect and runs

between the _____ and

_____ bones.

76 The calcaneofibular ligament joins the _____ to

the _____.

77 It is on the (medial/posterior/lateral) aspect of the ankle.

78 The medial, or *deltoid* ligament joins the tibia to the

_____ and _____ bones.

79 The true ankle joint is that between only the talus and the leg bones. The fit is indicated in Fig. 7-9.

80 The mortise-and-tenon type joint of the ankle between the

_____ , _____ , and

_____ is capable of movement in one plane only. The ankle joint permits only (inversion and eversion/plantar flexion and dorsiflexion/circumduction).

81 The remaining movements commonly thought to be those of the ankle are actually movements of the *intertarsal* joints. It is through these joints that the foot accommodates to uneven ground while one is hiking, etc.

82 The act of standing on a roof peak with one foot on either side of the ridge involves the (ankle joint/intertarsal joints).

inferior

navicular

calcaneus

fibula

calcaneus
lateral

calcaneus; navicular

fibula; tibia

talus
plantar flexion and
 dorsiflexion

intertarsal joints

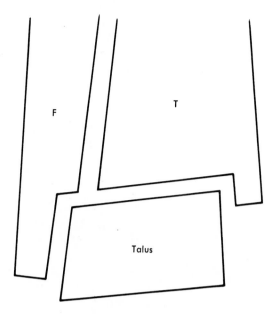

FIG. 7-9. The fit of the tibia and fibula over the talus creates the ankle joint.

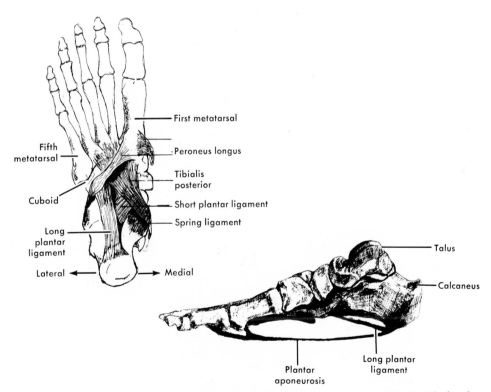

FIG. 7-10. Sole of the right foot. The long plantar ligament should not be confused with the plantar aponeurosis (the latter does, however, offer some support to the arch). The three ligaments, long, short, and spring ligaments, are also supported by wide tendinous insertions of some muscles such as the tibialis posterior shown here on the plantar view of a right foot.

101

83 The arches of the foot are described as the *medial* and *lateral longitudinal arches.* There is some doubt that a transverse arch across the heads of the metatarsals exists.

long plantar

84 Examine Fig. 7-10 and note that the lateral arch is supported by the (long plantar/spring/short plantar) ligament.

short plantar

85 The medial arch is supported by the (long plantar/short

spring

plantar) ligament and the _____ ligament.

86 The two muscles most important is assisting the ligaments to maintain the arches are a major invertor of the foot, the

posterior

tibialis _____ , and a major evertor of the

longus

foot, the peroneus _____ .

INTRINSIC MUSCLES OF THE FOOT

87 There are both extrinsic and intrinsic muscles of the foot. The extrinsic muscles have been dealt with. The intrinsic muscles have two essential origins:

> *calcaneus*—the tuberosity, or the undersurface
> *metatarsals*—the bases and along the shafts

intrinsic
calcaneous and meta-
tarsal

88 The (extrinsic/intrinsic) muscles of the foot originate on the (calcaneus/navicular/metatarsal/cuboid) bones.

89 There are four layers of muscle in the sole of the foot. From superficial to deep, that is, from the skin upward, they are the:

> "tie-beam" muscles
> flexor accessorius and four lumbricals
> short flexors of the first and fifth digits and adductor hallucis
> seven interossei

90 The first group of muscles (Fig. 7-11), known as the

tie-beam

_____ , muscles, consists of the abductor hallucis, the abductor digiti minimi, and the flexor digiti brevis.

calcaneus

91 All three muscles originate on the heel bone, or _____ .

92 The abductor hallucis inserts onto the base of the first phalanx of the big toe, and similarly, the abductor digiti

little toe, or
  fifth digit

minimi inserts onto the base of the first phalanx of the

_____ .

has not

93 The flexor digitorum brevis (has/has not) an analogous structure in the hand, and it inserts into the base of the

second; third

distal phalanges of the _____ , _____ ,

fourth; fifth

_____ , and _____ toes.

94 The term *tie-beam* refers to the function of these muscles to help "tie" into the extreme ends of the foot to support the

102

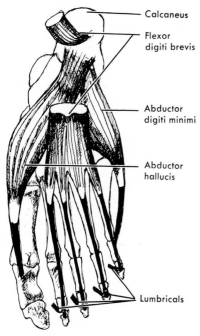

Calcaneus

Flexor digiti brevis

Abductor digiti minimi

Abductor hallucis

Lumbricals

Fig. 7-11

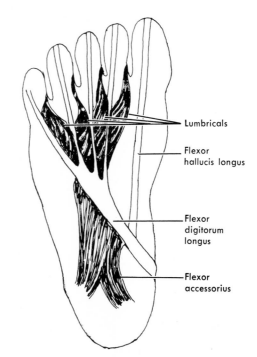

Lumbricals

Flexor hallucis longus

Flexor digitorum longus

Flexor accessorius

Fig. 7-12

FIG. 7-11. First layer (superficial) of muscles of the sole of the foot.
FIG. 7-12. Second layer of muscles of the sole of the foot.

arch. In addition, of course, they supply the functions their

abducts — names suggest: the abductor hallucis _____ the

abducts — big toe, the abductor digiti minimi _____ the

flexes — little toe, and the flexor digitorum brevis _____ the lateral four toes.

95 The muscles of this layer are innervated by branches of the tibial nerve, either the medial plantar or lateral plantar nerves. In this case the abductor hallucis would be inner-

medial — vated by the (medial/lateral) plantar nerve.

96 The flexor digitorum brevis and abductor digiti minimi are innervated by the other branch of the tibial nerve, the

lateral plantar — _____ nerve.

97 The second layer of muscles of the sole of the foot (Fig. 7-12) consists of a group of muscles analogous to those in

lumbricals — the hand known as the _____ , or wormlike muscles, and the flexor accessorius.

98 The flexor accessorius arises as did the muscles in the first

calcaneus — layer, from the _____ , but it inserts into the tendon of the flexor digitorum longus, an

extrinsic — (extrinsic/intrinsic) muscle of the foot.

103

flexor digitorum longus

99 The lumbricals arise from the tendon of the same muscle into which the flexor accessorius inserts, the _____

_____ muscle.

100 The lumbricals insert into the bases of the proximal phalanges of the second to fifth toes.

101 The job of the flexor accessorius longus is to supply a direct pull on the obliquely lying tendon of the flexor digitorium longus. To do this, it is innervated by the same nerve

lateral plantar

supplying the abductor digiti minimi, the _____

_____ nerve.

102 The lumbricals help flex the metatarsophalangeal joint, or M.P. joint. The lateral plantar nerve stimulates all but the

M.P.

first lumbrical in their action of flexing the _____ joint. The first lumbrical is innervated by the remaining

medial plantar

nerve of the sole of the foot, the _____ nerve.

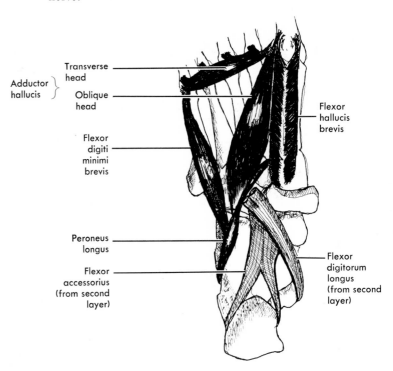

FIG. 7-13. Third layer of muscles of the sole of the foot.

103 The action of flexion of the lumbricals is aided by the muscles of the next layer (Fig. 7-13). These muscles are the flexor digiti minimi, the flexor hallucis brevis, and the adductor hallucis. Because of its medial position the innervation of the flexor hallucis brevis is by the

medial plantar

_____ nerve, whereas that of the flexor

lateral plantar

digiti minimi is by the _____ nerve.
The adductor hallucis is innervated by the lateral plantar
nerve.

104 The flexor digiti minimi originates on the fifth metatarsal
and inserts into the base of the proximal phalanx of the

fifth metatarsal

little toe. From its origin on the _____ the
flexor digiti minimi contracts to provide the action of

flexion

_____ of the little toe.

105 The flexor hallucis brevis originates on the intermediate and
lateral cuneiform and inserts as a counterpart to the flexor

proximal

digiti minimi onto the base of the _____
phalanx of the big toe.

106 The adductor hallucis brevis has an oblique and transverse
head. The oblique head arises from the tendon of the

transverse

peroneus longus, and the second, or _____ , head
arises from the ligaments of the M.P. joints.

107 The oblique head proceeds from its origin on the tendon of

peroneus longus

the _____ to insert in common

proximal phalanx

with the flexor hallucis brevis onto the _____

_____ .

108 The transverse head proceeds from its origin on the

ligaments of the M.P.
joints

_____ to insert on-
to the tendon of the flexor hallucis longus.

109 Thus the two heads of the adductor hallucis insert in
common with two flexors of the big toe, the oblique head

brevis

with the flexor hallucis _____ and the trans-

longus

verse head with the flexor hallucis _____ .

lateral plantar

110 The adductor hallucis is supplied by the _____

_____ nerve, and, as its name suggests, it

adducts

_____ the big toe.

111 The line to which the toes are adducted or from which the
toes are abducted runs through the center of the second toe.

fourth

Thus the last, or (fourth/third/fifth), set of muscles of the

interossei

foot, the _____ , are concerned about the
movements of abduction and adduction about this line

second

running through the _____ toe.

112 There are three plantar interossei arising from the medial

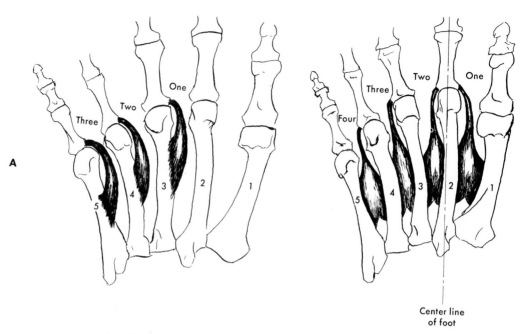

FIG. 7-14. Fourth layer of muscles of the foot—the interossei. **A,** Plantar. **B,** Dorsal.

side of the lateral three metatarsals. Each inserts into the medial side of the base of the proximal phalanx. Both plantar and dorsal interossei (Fig. 7-14) are innervated by the lateral plantar nerve.

113 A plantar interosseous muscle from its origin on the

metatarsal _____ bone and its insertion on

proximal

adduct

the base of the _____ phalanx must act to (abduct/adduct) the toes as well as flex the M.P. joint.

114 The dorsal interossei are four in number and arise from adjacent metatarsals; there is one in each intermetatarsal space. Note on Fig. 7-14, *B,* that the dorsal interossei all insert on the lateral side of the base of the first phalanx

first

except the _____ one.

115 From the position of the insertion of a dorsal interossei one

abduct

flex

could deduce that it would (adduct/abduct) the toe as well as (flex/extend) it under stimulation from the

lateral plantar _____ nerve.

116 The mnemonics for remembering the functions of the interossei of the *hand* are PAD (palmar adduct) and DAB

does

(dorsal abduct). This (does/does not) hold true for the interossei of the feet.

117 In the act of curling the toes under or plantar flexing them and the foot as a gymnast might do in "extending" his toes, there are a number of intrinsic and extrinsic muscles brought

four

into play. The intrinsic muscles would include, from deep to superficial, muscles in (one/two/three/four) layers.

118 The dorsum of the foot has one intrinsic muscle only, the extensor digitorum brevis, which arises from the upper surface of the calcaneus and inserts into the four medial toes. From its position the muscle probably helps the act of

extension

_____ of the four medial toes when stimulated by the deep peroneal nerve.

119 The blood supply to the lower limb is summarized in Fig. 7-15. The nerve supply to the lower limb is summarized in Fig. 7-16.

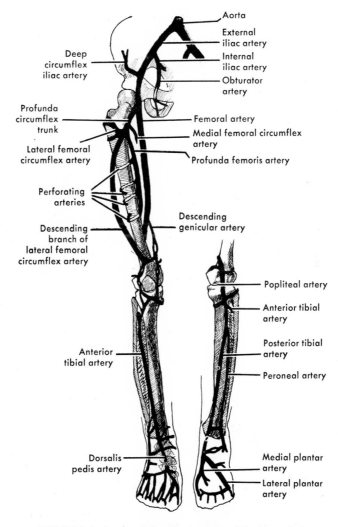

Aorta
External iliac artery
Internal iliac artery
Obturator artery
Deep circumflex iliac artery
Profunda circumflex trunk
Femoral artery
Medial femoral circumflex artery
Lateral femoral circumflex artery
Profunda femoris artery
Perforating arteries
Descending genicular artery
Descending branch of lateral femoral circumflex artery
Popliteal artery
Anterior tibial artery
Posterior tibial artery
Anterior tibial artery
Peroneal artery
Dorsalis pedis artery
Medial plantar artery
Lateral plantar artery

FIG. 7-15. Summary of the blood supply of the lower limb.

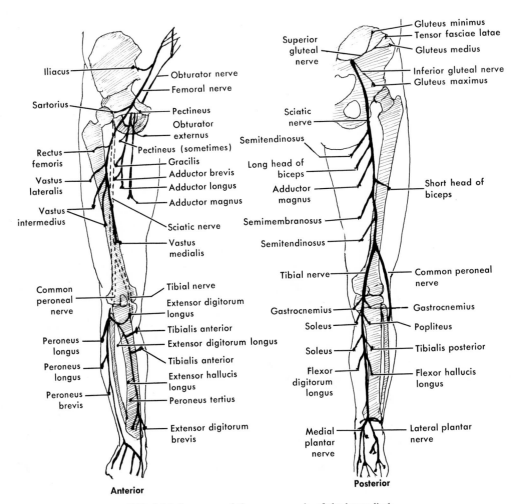

Iliacus

Sartorius

Rectus
femoris

Vastus
lateralis

Vastus
intermedius

Common
peroneal
nerve

Peroneus
longus

Peroneus
longus

Peroneus
brevis

Obturator nerve

Femoral nerve

Pectineus

Obturator
externus

Pectineus (sometimes)

Gracilis

Adductor brevis

Adductor longus

Adductor magnus

Sciatic nerve

Vastus
medialis

Tibial nerve

Extensor digitorum
longus

Tibialis anterior

Extensor digitorum longus

Tibialis anterior

Extensor hallucis
longus

Peroneus tertius

Extensor digitorum
brevis

Anterior

Superior
gluteal
nerve

Sciatic
nerve

Semitendinosus

Long head of
biceps

Adductor
magnus

Semimembranosus

Semitendinosus

Tibial nerve

Gastrocnemius

Soleus

Soleus

Flexor
digitorum
longus

Medial
plantar
nerve

Gluteus minimus

Tensor fasciae latae

Gluteus medius

Inferior gluteal nerve

Gluteus maximus

Short head of
biceps

Common peroneal
nerve

Gastrocnemius

Popliteus

Tibialis posterior

Flexor hallucis
longus

Lateral plantar
nerve

Posterior

FIG. 7-16. Summary of the nerve supply of the lower limb.

108

The vertebral column and back

1 The muscles of the back are attached directly or indirectly to the twenty four moveable bones of the vertebral column and to the immoveable *sacrum* and *coccyx*. The moveable bones are separated by discs of cartilage, whereas the immoveable

sacrum; coccyx _____ and _____ are not.

discs of cartilage **2** The moveable vertebrae separated by _____ are divided into three regions. The regions, from superior to inferior, are:

cervical (neck)
thoracic (trunk)
lumbar (lower back)

cervical **3** The neck, or _____ , vertebrae are seven in number. There are twelve thoracic vertebrae.

lumbar **4** The lowest portion of the moveable spine, the _____ , has five vertebrae, compared to the thoracic region's

twelve _____ . Thus the total of twenty-four moveable

seven vertebrae is made up of _____ cervical,

twelve; five _____ thoracic, and _____ lumbar vertebrae.

5 Examine Fig. 8-1 and note that the spine of a vertebra is the
posterior most (anterior/posterior) projection and that the direction of the spines changes from cervical to lumbar.

outward **6** The spines project (outward/downward) in the upper third of
downward the vertebral column and point (outward/downward) in the middle third of the column.

7 The vertebrae interlock in such fashion that they move on one another much as a gooseneck lampstem functions. The points of the vertebrae that articulate most are the lateral

transverse projections, called _____ *processes.*

8 The intervertebral discs are made of fibrocartilage and are necessary in part to give flexibility of the column and to absorb shock. Thus twisting the trunk is an example of

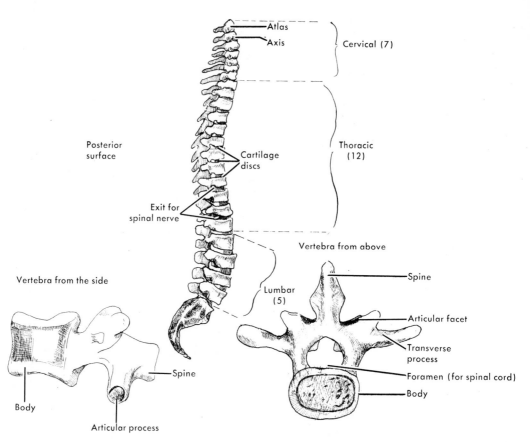

FIG. 8-1. Spinal column and a typical vertebra. Note the direction the spines of the vertebrae assume in the different areas.

flexibility	_____ of the spine due partly to the pres-
discs	ence of the _____ . Jumping tests the
shock-absorbing	_____ qualities of the discs.
	9 The discs receiving most shock in the landing after a jump for a rebound would be those with most weight on them, or
lumbar	those in the _____ region of the vertebral column.
	10 The lumbar vertebrae are also the thickest and heaviest for the same reason, that is, they carry most weight. The
cervical	vertebrae carrying least weight are the _____ vertebrae, which are also the lightest in structure.
	11 The first cervical vertebra is the *atlas*, which, of course,
skull	articulates with the _____ above it.
atlas	**12** The first vertebra, or the _____ , also articulates with the second vertebra, or the *axis*.

13 The joint between the skull and atlas is called the *atlanto-occipital joint,* and that between the atlas and axis is called

axial

the atlanto-_____ joint.

atlanto-occipital

14 The construction of the first, or _____ , joint between the atlas and occipital bone of the skull permits the nodding of the head for "yes."

atlantoaxial

15 The second, or _____ , joint permits shaking the head for "no."

axis

16 The second cervical vertebra, or _____ , has an upward projecting finger of bone called the *dens* (= tooth) that acts as a pivot point for the atlas.

dens

17 The presence of the toothlike _____ on the

axis

_____ explains why the atlantoaxial articulation affords the action of shaking the head for "no."

yes; no

18 In short, the atlas says (yes/no), and the axis says (yes/no).

19 The most prominent vertebra is T1 (thoracic vertebra 1). By flexing your neck and running your fingers down the cervical spines, you will easily palpate T1 projecting further than the

7

others. It lies just inferior to C _____ .

20 The deep muscles of the back lie in "gutters" on each side of the vertebral column. The *erector spinae* is the name of one group of muscles loosely divided into three columns lying in

gutters

the _____ on each side of the vertebral column.

erector spinae

21 The _____ muscles lying in the "gutters" are named from medial to lateral:

> spinalis
> longissimus thoracis
> iliocostalis

22 The erector spinae as a group originate on the sacrum and the crest of the ilium. Their insertions are varied, but their innervation is by the dorsal rami of the adjacent spinal nerves.

sacrum

23 The spinalis muscle from its origin on the _____

ilium

and the crest of the _____ inserts onto the tips of the spinous processes of the vertebrae—hence its name.

24 The middle column of the erector spinae is the

longissimus thoracis

_____ . This muscle inserts onto the transverse processes of the thoracic vertebrae and also onto the mastoid process of the skull just behind the ear.

25 The insertion of the longissimus, or long muscle, onto the

transverse; thoracic

_____ processes of the _____

vertebrae is paralleled by the insertion of the iliocostalis onto the angle of the ribs.

26 The iliocostalis (costal = ribs) not only inserts onto the

angle of the ribs
seven

_____ but also onto the transverse processes of all (five/six/seven) cervical vertebrae.

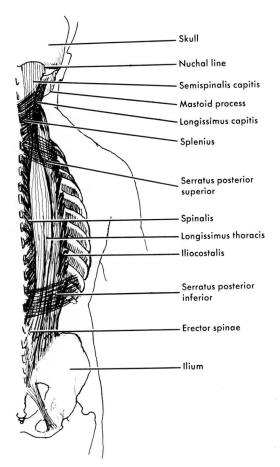

Skull

Nuchal line

Semispinalis capitis

Mastoid process

Longissimus capitis

Splenius

Serratus posterior superior

Spinalis

Longissimus thoracis

Iliocostalis

Serratus posterior inferior

Erector spinae

Ilium

FIG. 8-2. Erector spinae, splenius, and serratus posterior muscles.

27 The names of the three divisions of the erector spinae (Fig. 8-2) suggest something of their insertion: the spinalis to the

spinous processes

transverse processes

mastoid process

angle of the ribs

transverse processes

_____, the longissimus to the _____ of the thoracic vertebrae and as far as the _____, and the iliocostalis to the _____ and _____ of the cervical vertebrae.

28 The erector spinae are responsible for an action their name

erect

dorsal rami

suggests, that is, putting the spine into an ——————— position. This action is better known as extending the back. It is caused by stimulation from the ——————— of the spinal nerves.

29 The direction of the fibers of the erector spinae is upward and outward in a V shape. It can be called a *spinotransverse* system because the fibers run somewhat from the spinous processes to the transverse processes of the vertebrae. The

medial-lateral

fibers of the erector spinae therefore travel in a (lateral-medial/medial-lateral) direction.

30 The muscles beneath the erector spinae are shorter, run in the opposite direction (an inverted V) and are known conversely

transversospinalis
lateral-medial

as the ——————— system.

31 The transversospinalis muscles (Fig. 8-3) run in a (lateral-medial/medial-lateral) direction.

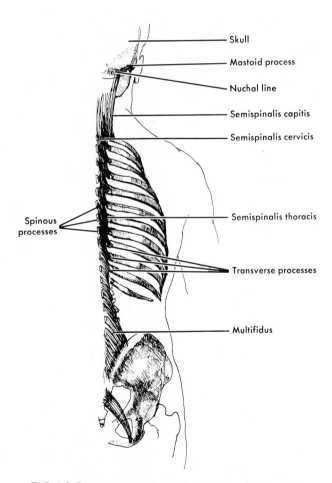

Skull

Mastoid process

Nuchal line

Semispinalis capitis

Semispinalis cervicis

Semispinalis thoracis

Spinous processes

Transverse processes

Multifidus

FIG. 8-3. Transversospinalis muscles (rotatores not shown).

113

32 The fibers of the transversospinalis group originate on the transverse processes of the vertebrae and insert on the

spinous
_____ processes of the vertebrae.

33 The function of the transversospinalis muscles is rather uncertain. They are divided into three lying from superficial

erector spinae
to deep rather than medial to lateral as the _____

_____ muscles are.

34 The transversospinalis muscles are from superficial to deep:

semispinalis (on the upper half of the vertebral column and covering two to six vertebrae)
multifidus (cover two to three vertebrae)
rotatores (cover one to two vertebrae)

35 The shortest muscles of the transversospinalis group are the

rotatores
_____ .

deep
36 The rotatores are the (intermediate/superficial/deep) layer of the transversospinalis.

37 The longest fibers in the transversospinalis group belong to

semispinalis; most
superficial
the _____ . It is the (deepest/most superficial) muscle of the group.

38 Two additional posterior muscles of the back, the *serratus posterior superior* and the serratus *posterior inferior,* lie superficial to the erector spinae (Fig. 8-2). They are so thin and weak that they have little functional significance.

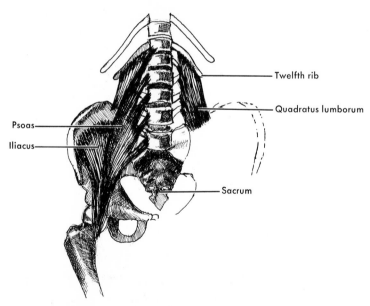

FIG. 8-4. Quadratus lumborum and psoas muscles. Although considered as part of the posterior wall of the abdomen, these muscles actually function on the vertebral column.

39 The *quadratus lumborum* (Fig. 8-4) runs from the iliac crest to the twelfth (last) rib and is deeply placed. Running from

iliac crest

its origin on the ———————————————— to the

twelfth

———————————— rib, the quadratus lumborum may be considered part of the posterior wall of the abdomen rather than a muscle of the back.

40 From its attachments one could deduce that the quadratus

extend; laterally flex

lumborum would act to (flex/extend/laterally flex) the spine.

41 Lateral flexion would occur when only one quadratus lumborum is contracted in conjunction with the erector

same

spinae on the (same/opposite) side.

42 The quadratus lumborum is innervated by the *subcostal* (beneath the rib) nerve and the *lumbar plexus.*

43 The act of bending sideways such as a gymnast or a dancer

quadratus lumborum

does is caused by the contraction of the ————————————

erector spinae

———————————— muscle and ———————————— muscles on one side of the back.

subcostal

44 The quadratus lumborum is innervated by the ——————————

lumbar plexus

and ———————————— nerves, and the erector

dorsal rami of the spinal nerves

spinae are innervated by the ————————————————

———————————— .

45 When lying prone and arching the back, the muscles most

erector spinae
dorsal rami of the spinal nerves

involved are the ————————————, which are innervated by the ————————————————————.

46 The more oblique the course of the muscle bundle, the more it is concerned with rotation. Thus the action of twisting the trunk while the feet are planted is made possible by the

transversospinalis

(transversospinalis/spinotransverse) muscles.

47 The action of lateral flexion automatically involves rotation as well. It would involve not only the transversospinalis

erector spinae

muscles but also the ———————————————— group

quadratus lumborum

and the ———————————— functioning together.

48 The rotation that accompanies lateral flexion may result from the gliding of the articular facets of the vertebrae on each other as well as from the presence of the intervertebral

discs

———————— .

49 Standing quietly does not involve the back muscles; however, in virtually every sport the back muscles play an important part.

50 In Fig. 8-5 note the relationships in cross section of the back muscles.

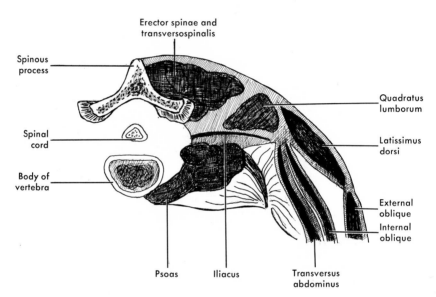

FIG. 8-5. Horizontal section through the lumbar area. Note the positions of the muscles relative to the vertebral column.

116

$\mathcal{9}$ The head and neck

PREVERTEBRAL MUSCLES

1 The muscles on the front (prevertebral surface) of the vertebral bodies are found only in the neck and lumbar region. These muscles on the (posterior/anterior/lateral) surface of the cervical vertebrae are the:

anterior

> longus capitis
> longus cervicis
> rectus capitis anterior
> rectus capitis lateralis

2 Examine Fig. 9-1 and note that the rectus capiti muscles

first cervical vertebra

extend to the head (caput) from the _____.

Rectus capitis
anterior

FIG. 9-1. Prevertebral muscles in the cervical region.

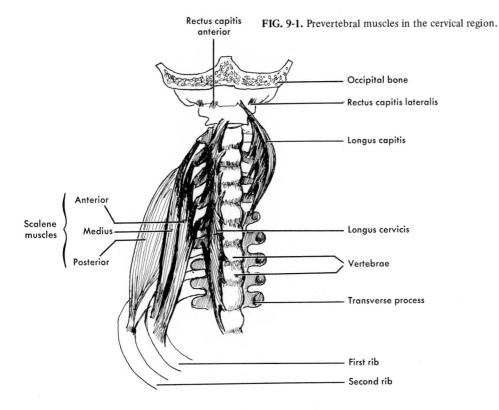

Occipital bone

Rectus capitis lateralis

Longus capitis

Scalene
muscles
Anterior

Medius

Posterior

Longus cervicis

Vertebrae

Transverse process

First rib

Second rib

3 The only other prevertebral muscle of this region running to

longus capitis
the head is the _____ , as the name suggests.

4 The longus cervicis originates on the upper thoracic vertebrae

cervical
and inserts onto the _____ vertebrae as its
name suggests.

cervicis
5 Thus there are two longus muscles, the longus _____

capitis
and the longus _____ , and two recti muscles, the

anterior; lateralis
rectus capitis _____ and _____ .

6 They are innervated by branches of the ventral (anterior) rami
of the cervical nerves; from their position one could deduce

flex
that the prevertebral muscles (flex/extend) the head.

7 The action of raising the head when lying supine is caused in

prevertebral
part by the (erector spinae/splenius/prevertebral) muscles

anterior
innervated by the (anterior/posterior) rami of the spinal
nerves.

8 In the lumbar region the prevertebral muscle is the psoas
muscle (Fig. 8-4), which is concerned primarily with

flexion
_____ of the hip.

9 Although it is a hip flexor, the psoas can act to flex the trunk
on the hip such as when performing a sit-up while the legs are
extended. Such an action pulls the lumbar spine forward and
can cause damage.

POSTERIOR NECK MUSCLES

10 The muscles of the back of the neck are somewhat distin-
guishable from the muscles of the back. The neck muscles are
closely associated with the ligamentum nuchae (= neck)
running from C7 to the skull.

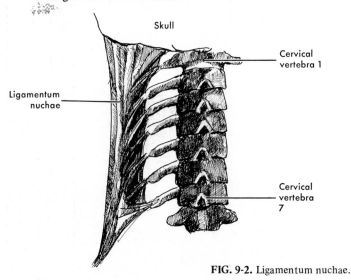

FIG. 9-2. Ligamentum nuchae.

seventh cervical vertebra; skull	**11** The ligamentum nuchae (Fig. 9-2) running from the _____ to the _____ acts as a muscle attachment.
	12 The several muscles attached to the ligamentum nuchae and to the skull fill in the back of the neck. The two most important are the *splenius* (bandage) and *semispinalis capitis*
splenius	(Fig. 8-2). These two muscles, the _____ , or
semispinalis capitis	bandage, and the _____ , originate on the cervical and upper thoracic vertebrae as well as
ligamentum nuchae	the _____ . The splenius lies deep to the semispinalis capitis.
beneath	**13** The splenius runs obliquely from its origin (beneath/above)
upper thoracic	the semispinalis capitis on the _____ and
cervical; ligamentum nuchae	_____ vertebrae and _____ to the mastoid process of the skull just behind the ear and the nuchal line of the skull.
spinotransverse	**14** The splenius may be termed a (transversospinalis/ spinotransverse) muscle.
	15 The semispinalis capitis inserts on the occipital bone of the skull medial to the splenius. The semispinalis capitis arises on
cervical	the transverse processes of the _____
thoracic	and _____ vertebrae and runs somewhat me-
transversospinalis	dially. Therefore it is termed a (spinotransverse/ transversospinalis) muscle.
	16 Judging from their positions, when they contract, the splenius
extend and rotate	and semispinalis serve to (flex/extend/rotate) the head.
splenius	**17** A swimmer performing the crawl uses his _____ and
semispinalis	_____ muscles to control his head while breathing.
	18 The splenius and semispinalis, like the muscles of the back,
dorsal	are innervated by the (dorsal/ventral) rami of the
spinal	_____ nerves.
	19 Palpate your own neck. Locate the median furrow. The bulge
semispinalis capitis	on either side marks the (splenei/the semispinalis capitis).
	20 Beneath the splenius and semispinalis are four small muscles in a region known as the *suboccipital* region (Fig. 9-3). The
suboccipital	four muscles of the _____ region are the:

      rectus capitis posterior major
      rectus capitis posterior minor
      obliquus capitis inferior
      obliquus capitis superior

119

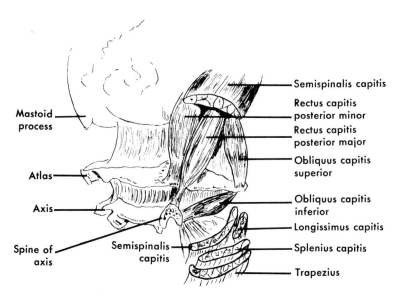

FIG. 9-3. Suboccipital region. The many small muscles illustrated here, plus the ligamentum nuchae, virtually fill the concavity created by the normal cervical curve of the vertebral column.

pairs

21 The suboccipital muscles are in (pairs/threes/fours).

22 A glance at Fig. 9-2 shows that the occipital bone of the skull protudes posteriorly a good distance further than the spines of the cervical vertebrae. The recti capitii posteriors originate on the spines of the atlas and axis and insert onto the nuchal

upward

line. The fibers therefore run in an _____ and

backward

_____ direction.

23 The recti capitii posteriors running from their origin on the

atlas; axis

spines of the _____ and _____ vertebrae are accompanied by the obliquus capitis inferior and the obliquus capitis superior.

24 The obliquus capitis inferior originates on the spine of the

2

axis, or C_____, whereas the obliquus capitis superior originates on the transverse process of the atlas.

25 The inferior oblique runs from its origin on the axis to the

1

transverse process of the atlas, or C _____ , and the

transverse process

superior from its origin on the _____ of the atlas to the occipital bone.

26 The obliquus capitis superior is almost a continuation of the inferior in that its origin is the insertion of the other.

27 The only muscle of the four not inserted on the occipital

obliquus capitis inferior

bone is the _____ .

120

28 The innervation of the suboccipital muscles is by the suboccipital nerve.

29 The action of the suboccipital muscles is to aid in rotation and extension of the head; thus, when a wrestler performs a "bridge," he utilizes (flexors/extensors) of the head and neck.

extensors

30 The muscles acting on the head in a bridge are (apart from the trapezius) the _____ , which is most superficial,

splenius

erector spinae

and the next deeper set of muscles, the _____ .

31 The third set of muscles acting on the head during a bridge is

semispinalis capitis

the _____ .

32 Beneath the semispinalis are the deepest muscles of the neck

suboccipital muscles

region, the _____ .

33 The muscles or sets of neck muscles used in bridging are

dorsal

innervated by (ventral/dorsal/lateral) spinal nerves except for the suboccipital muscles, which are supplied by the

suboccipital

_____ nerve.

OTHER NECK MUSCLES

34 The muscles of the neck remaining to be studied are the:

scalene (three in number; scalene = uneven)
sternocleidomastoid (from the sternum, or breastbone)
infrahyoid (below the hyoid bone in the throat)

three

35 The scalene muscles (Fig. 9-1) in the neck are (four/two/three) in number.

36 The sternocleidomastoid (Fig. 9-4) runs from the sternum, or

breastbone

_____ , to the mastoid process of the skull.

37 The infrahyoid muscles (Fig. 9-5) are found below the

hyoid

_____ bone in the throat.

38 The three scalene muscles (Fig. 9-1) originate on the transverse processes of all cervical vertebrae, giving them

seven

(five/six/seven) points of origin.

transverse processes of
the cervical vertebrae

39 From their origin on the _____

_____ the scalene muscles insert on-to the first and second ribs.

40 From their origins on the transverse processes of the cervical vertebrae and their insertions on the first and second ribs, one

flex

could say that they (flex/extend) the cervical part of the vertebral column.

41 The scalene muscles are also important during breathing and coughing. They are innervated by the ventral rami of the cervical spinal nerves.

42 The most important flexor of the head is the sternocleido-

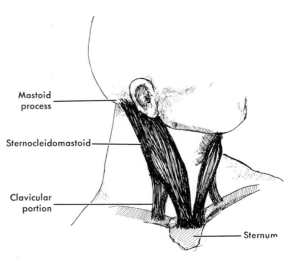

Mastoid process

Sternocleidomastoid

Clavicular portion

Sternum

FIG. 9-4. The sternocleidomastoid muscle shown turning the head.

mastoid (Fig. 9-4), arising from the breastbone, or

sternum   _____ , and often the clavicle as well.

43 The sternocleidomastoid inserts onto the mastoid bone and is innervated by the *accessory* nerve. Thus when it contracts

flex   (both sides simultaneously), it will _____ the head.

accessory   **44** When the _____ nerve stimulates the sternocleidomastoid on one side only, the head turns.

45 Palpate each of your sternocleidomastoid muscles while turning your chin hard to the right and back a few times. The

left   sternocleidomastoid doing the work is the (right/left) muscle.

46 When performing a sit-up, one raises his head by the

sternocleidomastoid   _____ muscles on stimulation by the

accessory   _____ nerve.

47 The blood supply to the entire area is from branches of the external carotid artery on each side.

48 The *hyoid* bone is situated in the angle where the floor of the mouth meets the front of the neck. The hyoid is therefore

prevertebrally   located (postvertebrally/prevertebrally).

49 The hyoid provides attachment for a series of muscles called

below   the *infrahyoid* muscles, which lie (above/below) the hyoid bone.

50 The infrahyoid muscles are four in number, and they run,

hyoid   with one exception, from the sternum to the _____ bone (Fig. 9-5).

four   **51** The (three/two/four) infrahyoid muscles are the:

       sternohyoid   omohyoid   sternothyroid   thyrohyoid

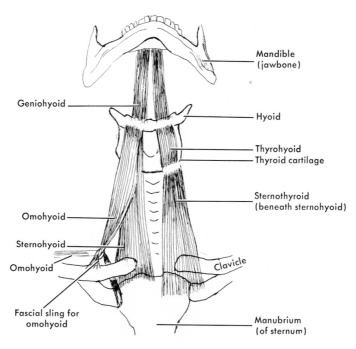

FIG. 9-5. Infrahyoid muscles.

52 The sternohyoid, as its name suggests, originates on the

sternum; hyoid _____ and inserts on the _____ .

53 The omohyoid originates on the upper border of the scapula

hyoid near the suprascapular notch and inserts on the _____ .
It is tied down about halfway by a fascial sling to the
sternum and clavicle. This allows the second part of the

omohyoid _____ muscle to change direction from
horizontal to vertical.

54 The sternothyroid muscle arises on the sternum and inserts

thyroid onto the _____ .

55 The thyroid cartilage is part of the *larynx*, or voice box, and
it is the origin of the thyrohyoid, as well as the insertion of

sternothyroid the _____ .

thyrohyoid **56** The last muscle of the four, the _____ ,
also inserts on the hyoid bone after leaving its origin on the

thyroid cartilage _____ .

57 The infrahyoid muscles all run in a vertical direction except

omohyoid for the _____ muscle.

voice box **58** The infrahyoid muscles depress the larynx, or _____

_____ , and act as protection for the organs
behind them. The infrahyoid muscles are innervated from the

123

ventral

same source as other muscles in this region of the neck, the (dorsal/ventral) rami of the spinal nerves.

INTERNAL STRUCTURES OF THE NECK

59 The internal structures of the neck are many, including the following:

> trachea
> esophagus
> larynx
> carotid arteries
> nerves
> endocrine glands

anterior

60 Note on Fig. 9-6 that the trachea is (anterior/posterior) to the esophagus.

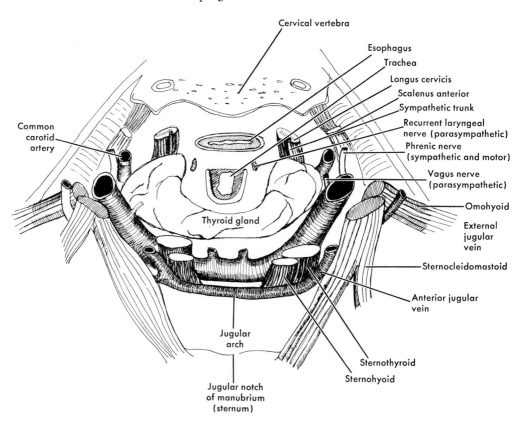

FIG. 9-6. Horizontal section through the base of the neck.

61 The trachea, or windpipe, leads to the lungs and is reinforced by half-rings of cartilage to ensure that the passage remains

trachea

open. The windpipe, or _____ is about 1 foot long.

larynx

62 The voice box, or _____ , is a part of the trachea.

124

63 Situated behind the trachea is the esophagus, which leads to the stomach. The esophagus, which is very distensible, lies

vertebrae

just anterior to the _____.

64 The carotid arteries are the major blood supply to the head and brain. They lie on either side of the trachea. The

carotid

_____ arteries to the brain are protected by the more superficial muscles.

65 Several nerves in the neck are major suppliers of *autonomic* (see discussion in Chapter 17) impulses to other parts of the body. The nerves in the area are the *phrenic, vagus,* and

autonomic

recurrent laryngeal nerves, all of which carry _____ impulses.

66 The thyroid gland is an endocrine gland lying on the anterior

thyroid

surface of the trachea. The _____ gland does

trachea

not extend the whole length of the _____ , or windpipe, but is confined to the upper portion in an H shape across the windpipe.

67 The thyroid gland controls body metabolism. It has a smaller gland situated in each end of each leg of the H called the *parathyroid* gland. Unlike the thyroid, which controls

metabolism

_____ , the parathyroid controls the level of calcium in the blood.

thyroid

68 The two glands, the _____ gland and the

parathyroid

_____ gland, lie posterior to the left brachiocephalic vein at about the level of the jugular notch of the sternum.

69 The jugular veins empty into the brachiocephalic veins, which lead to the heart.

70 A pulse may be distinctly felt if you palpate about 4 cm.

common carotid

above the sternoclavicular joint. The _____ artery at this point may be compressed against the transverse processes of the cervical vertebrae.

10 The thorax and abdomen

MUSCLES OF THE THORAX

1 The thorax refers to the part of the body enclosed by the ribs and is therefore that part of the body between the (neck and waist/waist and hips/neck and lowest rib).

neck and lowest rib

2 Examine Fig. 10-1 and note that there are (six/twelve/ten) ribs.

twelve

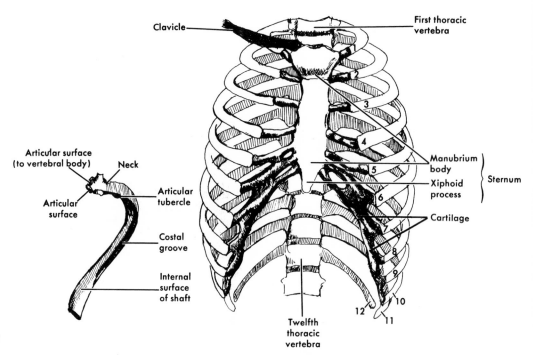

FIG. 10-1. The rib cage, front view, and a typical rib. Note that ribs 11 and 12 are "floating" ribs. They are not attached to the sternum. A typical rib turns downward and twists on its way to joining the sternum.

3 The ribs run from the vertebral column forward to the _____ .

sternum

4 The ribs are joined to the sternum by (ligament/tendon/cartilage).

cartilage

126

5 The muscles of the thorax are connected with protection of vital organs and the process of respiration (breathing).

6 The muscles of the thorax that are involved in the process of

breathing   _____ are the:

    intercostals (between the ribs)
    diaphragm
    levatores costarum

between

7 The intercostals run (behind/in front of/between) the ribs, and there are two layers of them, an internal and external set.

external

8 The outermost, or _____ , intercostals originate on the lower margins of the first eleven ribs and pass down and forward to the ribs below.

9 Placing your hand in your pant's pocket will indicate the direction of the fibers of the external intercostals. The internal costals originate from the same site, that is, the

lower margin of the
  first eleven ribs

_____ ,

and run down and backward to the same insertion.

10 Thus, with the external intercostals running down and

forward   _____ and the internal intercostals running

backward   down and _____ , there is a shutterlike appearance to the muscles.

11 The intercostals are important in the inspiration phase of breathing and are innervated by the intercostal nerves.

inspiration   During the _____ phase of breathing the intercostals contract and raise the ribs when stimulated by

intercostal   the _____ nerves.

12 The intercostal nerves and blood vessels travel in grooves (Fig. 10-1) on the lower margins of the ribs and are thus in

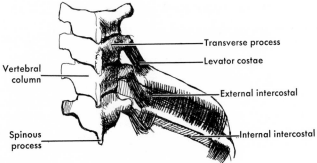

Transverse process
Levator costae
External intercostal
Internal intercostal
Vertebral column
Spinous process

FIG. 10-2. Posterior view of ribs attaching to the vertebral column. A rib articulates with the intervertebral disc and the vertebral body above and below the disc. The exceptions are ribs 11 and 12. The levator costae muscle and intercostal muscles are shown here.

origins

inspiration

levatores costarum

C7; T11

inspiration

dome

diaphragm
three

close proximity to the (origins/insertions) of the intercostal muscles.

13 The intercostal muscles have a secondary function in helping bend the trunk laterally. Their primary function in the _____ phase of breathing is assisted by the levatores costarum (Fig. 10-2).

14 The muscles assisting the intercostals in breathing, the _____ , arise on the transverse processes of vertebrae C7 to T11.

15 From their origins on the transverse processes of vertebrae _____ to _____ , the levatores costarum insert on the next lower rib.

16 The levatores costarum are innervated by the dorsal rami of the spinal nerves when they contract to assist the intercostals in (inspiration/lateral flexion).

17 The diaphragm (Fig. 10-3) is a dome-shaped sheet of muscle separating the thoracic and abdominal cavities. This _____-shaped muscle is attached completely around the inside of the body cavity, to the vertebral column, the last six costal cartilages and ribs, and the sternum. Thus the _____ is attached at (two/four/three) specific areas around its periphery.

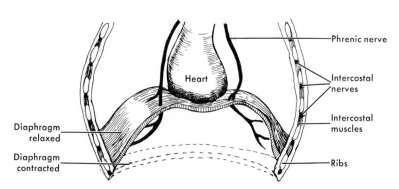

FIG. 10-3. Diaphragm in coronal section. Note that when the diaphragm is relaxed, it is up in a domed position.

18 The diaphragm has a central tendon on which the fibers pull. When the muscle contracts, the dome flattens, thus enlarging the thoracic cavity containing the lungs.

19 When the diaphragm contracts, the thoracic cavity gets (larger/smaller). In this activity it is assisted by the intercostals and levatores costarum.

larger

20 The diaphragm is supplied by the phrenic (diaphragm) nerve, which is a nerve of the autonomic nervous system. When the

phrenic
flattens
_____ nerve stimulates the diaphragm, the dome of the muscle (flattens/rises).

21 The diaphragm reaches as high as the sixth rib when you are standing at rest and the lungs are (full/empty).

empty

MUSCLES OF THE ABDOMEN

22 The anterior abdominal wall is designed to give great variation in the size of the abdominal cavity as the muscles comprising the anterior wall contract or relax.

23 The muscles of the anterior wall are the:

> external oblique
> internal oblique
> transversus abdominis
> rectus abdominis

Prepare a mnemonic for remembering these muscles.

five
24 There are (five/three/four) muscles comprising the anterior abdominal wall.

25 In addition to the muscles, the fascia of the abdomen is important. Thus there are two types of tissue important in

muscle
the anterior abdominal wall, _____ tissue and

fascia
_____.

26 The fascia is of special importance in the abdominal wall because *hernias* are associated with weakness in the fascia and muscle.

27 One of the problems of the abdominal wall is that of

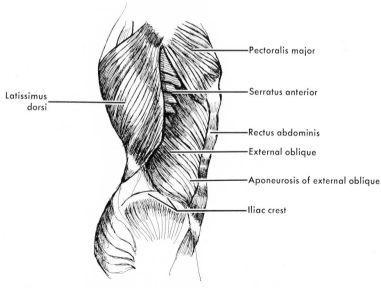

Latissimus dorsi

Pectoralis major

Serratus anterior

Rectus abdominis

External oblique

Aponeurosis of external oblique

Iliac crest

FIG. 10-4. Lateral view of the body showing the large triangular latissimus dorsi of the back partly overlaying the external oblique.

hernia

_____ brought on by weakness in the fascia and muscle. This condition often must be treated by surgery and may be brought on by the strain of unaccustomed exercise.

28 The external oblique is the most superficial of the antero-lateral muscles.

29 The external oblique (Fig. 10-4) is a muscle not only of the

lateral

anterior abdominal wall but also of the _____ abdominal wall.

30 The external oblique muscle originates on the lower eight ribs.

lower eight ribs

31 From its origin on the _____ the external oblique inserts on the *linea alba* (white line) (Fig. 10-5) and on the iliac crest.

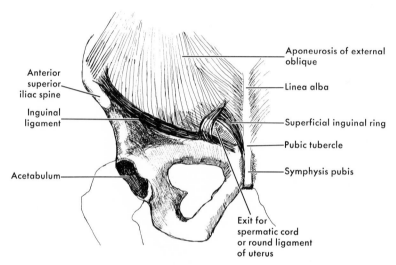

FIG. 10-5. The inguinal ligament is formed from the inferior edge of the external oblique.

32 The insertion of the external oblique on the white line, or

linea alba

_____ , brings it to the midline of the body, since the white line is a line of fascia down the center of the abdomen.

33 The fibers of the external oblique run in the same direction

downward

as the external intercostals, (upward/downward) and

forward

_____ .

34 The inferior edge of the external oblique leads from the anterior superior iliac spine to the symphysis pubis to form the *inguinal* ligament of the abdomen.

anterior superior

35 The inguinal ligament running from the _____

symphysis pubis

_____ iliac spine to the _____

130

_____ provides part of the muscle attachment for the internal oblique muscle.

36 Just above the medial end of the inguinal ligament there is a

external

split in the aponeurosis of the _____ oblique muscle called the *superficial inguinal ring.*

superficial inguinal
ring

37 The split in the aponeurosis called the _____

_____ permits the passage of the *spermatic* cord in males and the *round ligament of the uterus* in females.

spermatic cord
round ligament of the
uterus
internal

38 The passage of the _____ in males or

the _____

in females is medial to the origin of the (external/internal) oblique muscle on the inguinal ligament.

39 The internal oblique (Fig. 10-4) also originates on the anterior superior iliac spine and the thoracolumbar fascia. From these origins, it inserts with the external oblique into the

linea alba

_____ at the midline of the body.

inguinal ligament

40 After it leaves its origin on the (a) _____

iliac crest

_____ , (b) _____ , and

thoracolumbar fascia

(c) _____ , the internal oblique, which is the thickest muscle in the abdomen, becomes aponeurotic just before reaching the linea alba. Here, it splits to enclose the *rectus abdominis* muscle.

rectus abdominis

41 In splitting to enclose the _____ muscle, the internal oblique is somewhat incomplete because in its lower one fourth the aponeurosis passes only in front of, not behind, the muscle it encloses.

42 The next deepest muscle is the transversus abdominis, whose

horizontally

name implies that its fibers run (vertically/horizontally) on the abdominal wall.

43 The transversus is the deepest muscle of the abdominal wall, and it originates with the interal oblique on the

inguinal

_____ ligament and also on the iliac crest and thoracolumbar fascia.

iliac crest
thoracolumbar fascia

44 From its origin on the inguinal ligament, _____

_____ , and _____ ,

the transversus inserts with the aponeurosis of the internal

rectus abdominus

oblique to enclose the _____ muscle.

45 The rectus abdominus originates on the last segment of the

sternum

_____ , or breastbone, known as the *xiphoid* process.

131

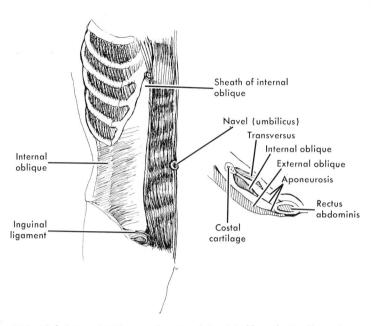

FIG. 10-6. Internal oblique and rectus abdominis. Note the tendinous intersection or interruptions in the fibers of the rectus abdominis. *Inset,* the aponeuroses of the internal oblique and transversus enclose the rectus abdominis.

xiphoid

46 From its origin on the _____ process of the sternum, the rectus abdominis (Fig. 10-6) runs straight down on each side of the linea alba to insert on the pubic symphysis.

47 The rectus abdominis, before reaching its insertion on the

symphysis pubis

_____, is marked by three or more tendinous intersections giving it, on a well-developed man, a "washboard" effect.

48 The four muscles of the anteroabdominal wall, the

external oblique; internal oblique; transversus abdominis; rectus abdominis; four

_____ , _____

_____ , _____ , and

_____ , run in (three/two/four) different directions.

49 The different directions of the four sets of muscle fibers plus the fascia of the anterolateral abdominal wall serve not only to provide movement of the trunk and maintenance of posture, but also to protect the viscera and increase intra-abdominal pressure.

trunk

50 In their several roles of movement of the _____ ,

viscera

maintenance of posture, protection of the _____ ,

intra-abdominal

and increasing _____ pressure, the anterolateral muscles of the abdomen are innervated by the intercostal nerves T7 to T11.

132

51 The control of intra-abdominal pressure by the muscles

intercostal

when stimulated by the _____

7; 11

nerves T _____ to T _____ is important in respiration, defecation, parturition, and vomiting.

52 The movement of the trunk accomplished by the antero-

flexion

lateral muscles is _____ at the hip in the supine position.

53 In addition to flexion, the anterolateral muscles are also responsible for lateral flexion. Thus in doing a sit-up with trunk twisting, the following anterolateral muscles of the

external oblique; inter-
nal oblique; rectus
abdominis; transver-
sus abdominis

abdominal wall are used: the _____ ,

_____ , _____

_____ , and _____

_____ .

CONTENTS OF THE ABDOMEN

54 The principal viscera of the abdomen are the:

      stomach
      liver and gallbladder
      intestines
      pancreas
      spleen
      suprarenal glands (or adrenal glands)
      kidneys
      ureters

55 Of the eight organs, the pancreas and three others are concerned with digestion. The three others are the

stomach; intestines

_____ , _____ , and

liver

_____ .

56 The kidneys and ureters are concerned with excretion of

urine, or liquid waste

_____ . The adrenal glands are part of the endocrine, or hormone system, and the spleen is a part of the circulatory system.

57 Apart from digestion, the following body systems are

excretory

represented by organs in the abdomen: the _____

endocrine

system, the _____ system, and the

circulatory

_____ system.

58 The stomach and intestines are attached to the body wall (Fig. 10-7) by folds of *mesentery* formed from the *peritoneum* lining the cavity. The kidneys, suprarenals, and pancreas are *retroperitoneal* (behind the peritoneum). The stomach and

anterior

intestines are thus (anterior/posterior) to the kidneys.

133

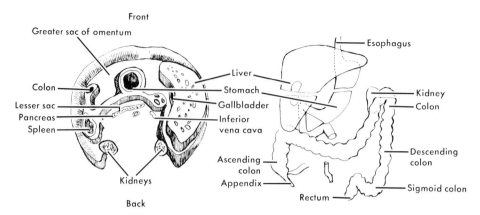

Front

Greater sac of omentum

Colon

Lesser sac
Pancreas
Spleen

Kidneys

Back

Liver
Stomach
Gallbladder
Inferior
vena cava

Ascending
colon
Appendix

Rectum

Esophagus

Kidney
Colon

Descending
colon

Sigmoid colon

FIG. 10-7. *Left,* horizontal section through the abdomen, at the stomach. *Right,* coronal section showing the same organs.

mesentery	**59** Folds of the peritoneum called _____ attach the stomach to the body wall.
	60 To understand how the peritoneum can line the inside of the body wall and cover the organs, think of a semi-inflated balloon. If one pokes a finger into the balloon, that finger becomes coated with a layer of rubber. Similarly, the viscera grow into the peritoneum from the body wall and receive a coating of peritoneum.
	61 The kidneys and their ureters are not covered with peri-
retroperitoneal	toneum and so are termed _____ .
are not	**62** The liver and stomach (are/are not) retroperitoneal.
	63 The peritoneum on the inside wall of the abdominal cavity is termed *parietal peritoneum,* and that on the viscera is termed the visceral peritoneum. The latter types of peritoneum may have specialized names such as *omentum* (a wide sheet of peritoneum).
	64 The peritoneum thus has two major divisions. That on the
parietal	walls is termed _____ and that on the
visceral	organs _____ peritoneum.
	65 Some types of visceral peritoneum have specialized names. A
omentum	broad sheet is called an _____ .
	66 The organs that have pushed into the peritoneum and thus have a wrapping have their nerve and blood supply conducted through the mesentery formed when they grew away from the abdominal wall. The mesentery also serves to combat infection and inflammation as well as conducting
nerves; blood vessels	the _____ and _____ to the organs.
	67 To help perform the part of its job concerned with

134

infection; inflammation

combating _____ and _____,
the mesentery of the greater curvature of the stomach (the
lower border) hangs like an apron over the small intestine.
This wide sheet of mesentery is called the greater

omentum

_____ and is invested with tags of fat.

68 The greater omentum hangs down in front of the

small intestine

_____ and is characterized by tags

fat

of _____ .

below

69 Examine Fig. 10-8 and note that the stomach, situated
(above/below) the liver, varies greatly in size and shape but
usually has a maximum length of about 10 inches and a
capacity of about 1 quart.

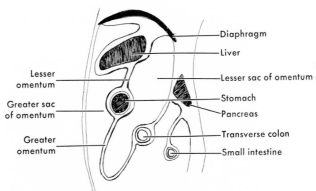

FIG. 10-8. Sagittal section showing the greater and lesser omentum.

1; 10

70 With a capacity of _____ quart(s) and a length of about _____
inches the variable-shaped stomach has a *cardiac* orifice at
its upper end and a *pyloric* (= gatekeeper) orifice at its lower.

71 The esophagus leads into the stomach at the upper, or

cardiac

_____ , orifice, and the small intestine
(Fig. 10-9) leads out of the stomach at the lower, or

pyloric

_____ , orifice.

72 Absorption through the stomach walls is slight. The main
function of the stomach is to provide enzymatic action to
break down food. The enzymes necessary come from glands
in the *mucosa* lining the stomach. Thus the prime job of the

enzymes

stomach is the secretion of _____ from glands in

mucosa

the _____ .

73 The pyloric *sphincter* (Fig. 10-10), a ring of muscle at the

pyloric

_____ orifice of the stomach, is thought

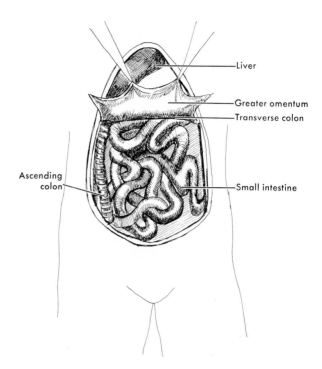

FIG. 10-9. Small and large intestines lying behind the flap of the greater omentum.

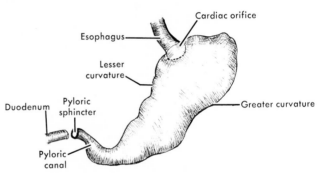

FIG. 10-10. Stomach. The shape is extremely variable.

to prevent the reflux (return) of matter from the duodenum, the next part of the digestive system.

74 Thus the food leaves the stomach and enters the next part of

duodenum

the digestive system, the _____ , which is about 10 inches long.

75 The duodenum (Fig. 10-11) is the beginning of the small intestine and receives the stomach contents, or *chyme*, first. The next section of the small intestine is the *jejunum;* the

chyme

_____ , or stomach contents, are passed from

duodenum; jejunum

stomach to _____ to _____ .

136

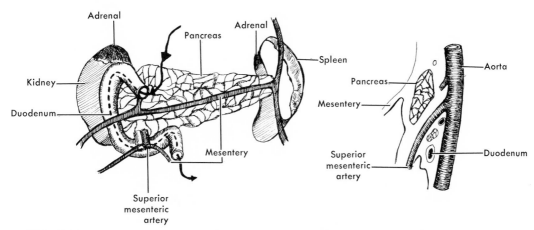

FIG. 10-11. *Left,* anterior view of duodenum and pancreas. *Right,* lateral view showing the pancreas as retroperitoneal.

76 The last section of the small intestine is the *ileum*, bringing the total length of the small intestine (including stomach) to about 21 feet. The last section of the small intestine, the

ileum; jejunum
_____ , lies between the _____ and the large intestine.

10
77 The duodenum is bent along its _____-inch length into a horseshoe shape, and the pancreas lies within the bend.

pancreas
78 The _____ lying within the bend of the duodenum secretes digestive juices into the duodenum.

79 In addition to the juices that it receives from the

pancreas
_____ , the duodenum also receives bile from the gallbladder and supplies digestive juices of its own.

80 The duodenum thus has digestive enzymes from these three

pancreas
sources to act on its contents: the _____ , the

gallbladder; duodenum
_____ , and the _____ .

81 The jejunum (= empty) is about 8 feet long. In the jejunum and ileum the food is digested and absorbed. The ends of the

posterior
small intestine are anchored by mesentery to the (anterior/ posterior) wall at the *duodenojejunal* junction and the *ileocecal* (cecum = blind) junction.

82 Apart from where it is anchored to the wall at the

duodenojejunal; ileo-cecal
_____ and _____ junctions, the small intestine is mobile and free to take many shapes and positions.

83 Within the small intestine the walls project with *villi* (= hairlike), giving the inner walls the texture of velvet. These

villi
hairlike _____ increase the area for absorption of digested food.

84 Each villus, to enhance its absorbing power, contains a small artery, vein, and lacteal (lymph vessel).

85 Examine Fig. 10-12 and note that in addition to the

artery; vein; lacteal

_____ , _____ , and _____ contained within each villus, there are also intestinal glands present in the walls of the small intestine.

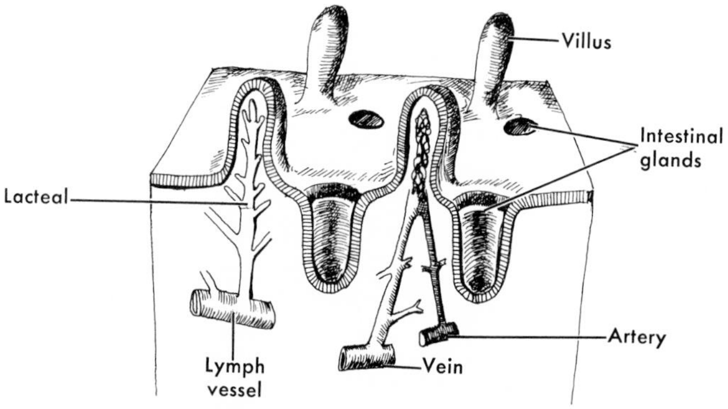

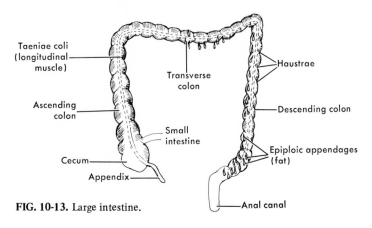

FIG. 10-12. Villi lining the inside of the small intestine.

blind

86 The large intestine consists of the cecum, or _____ section, and the *colon*. The ileum enters the cecum via the ileocecal valve. The larger section is the colon, and the smaller is the cecum.

87 The cecum lies in the right iliac fossa and forms a blind

ileum

pouch for the reception of the contents of the _____ .

right

88 The cecum lying in the (right/left) iliac fossa has attached to it a vermiform (= wormlike) *appendix*. The appendix, which is a blind pouch also, sometimes becomes inflamed and must be removed.

89 The colon comprises the major part of the large intestine

Taeniae coli (longitudinal muscle)

Transverse colon

Ascending colon

Small intestine

Cecum

Appendix

Haustrae

Descending colon

Epiploic appendages (fat)

Anal canal

FIG. 10-13. Large intestine.

138

and is divided into ascending, transverse, descending, and *sigmoid* sections, which gradually become narrower. The

sigmoid

narrowest portion, the _____ , ends at the *anus.*

90 Examine Fig. 10-13 and note that the large intestine is marked by *haustra,* or puckerings, which are caused by the pull of longitudinal muscle fibers. These puckerings, or

haustra

_____, create sacculations that help slow the passage of the contents through the large intestine.

91 The longitudinal muscles called *taeniae* (= tape) *coli* pull on

ascending

all four parts of the colon: the _____ ,

transverse; descending

_____ , _____ , and

sigmoid

_____ .

taeniae coli

92 The haustra created by the pull of the _____ muscles slows the passage of the waste products of digestion through the large intestine so that moisture may be absorbed from the waste, or *feces.*

feces

93 The solid waste, or _____ , is held in the anal canal by the *anal sphincter* until defecation or expulsion.

anal

94 The _____ sphincter that guards the feces from loss is controlled by nerves from the autonomic nervous system. There are also sensory nerves that register pain of distention but not of cutting or burning in the colon.

autonomic

95 The _____ nervous system not only controls the anal sphincter but also, in various forms, all the digestive tract. The nerves are distributed from *plexuses,* or relay stations, of which there are three major ones.

plexuses

96 The three major _____ , or relay stations, are the *celiac,* the *superior mesenteric,* and the

plexuses

inferior mesenteric _____ .

celiac; superior mesenteric; inferior mesenteric

97 The fibers from the _____ plexus, the _____ _____plexus, and the _____ _____ plexus travel largely along the arteries.

98 The arteries with which these autonomic fibers are associated are branches of the abdominal *aorta.* The names given

aorta

the major branches of the abdominal _____ are

celiac; superior mesenteric; inferior mesenteric

the same as those given the plexuses, the _____ , _____ , and _____ _____ arteries.

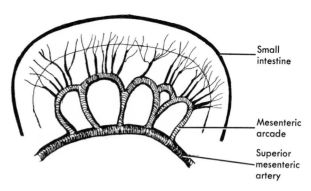

Small
intestine

Mesenteric
arcade

Superior
mesenteric
artery

FIG. 10-14. The mesenteric arcades ensure a generous blood supply to the intestines.

99 The celiac trunk, or artery, supplies the stomach and duodenum, and the two mesenteric arteries supply the rest of the small intestine. The mesenteric arteries form *arcades* (Fig. 10-14) to ensure a good blood supply to the last two

jejunum   segments of the small intestine, the _____ and

ileum   the _____ .

100 When the blood leaves the capillaries at the end of the

arcades   mesenteric _____ in the small intestine, it does not go as one would expect directly into veins leading back to the heart.

101 The venous blood from the small intestine goes to the liver by way of the *portal* vein to deliver its load of food

villi   absorbed through the _____ of the small intestine.

102 The venous blood carrying the food to the liver via the

portal   _____ vein then leaves the liver and proceeds back to the heart.

11 The abdominal viscera: liver, pancreas, and spleen

1 The abdominal viscera for purposes of study will be limited

spleen

here to the liver, pancreas, and spleen. The _____ is the only one of the three not connected with the digestive system. It is part of the circulatory system.

2 The liver (Fig. 11-1) is the largest organ in the body, weighing about 1500 grams, and lies in the thoracic cage covered by

superior

the diaphragm. It is thus (superior/inferior/lateral) to the stomach.

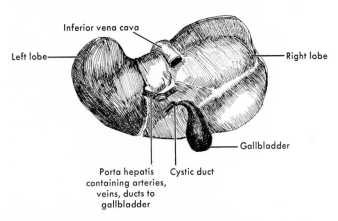

FIG. 11-1. View of the liver from below. Note that the gallbladder is on the inferior aspect.

mesentery

3 The lesser omentum, a sheet of _____, joins the liver to the stomach.

4 The function of the liver is to secrete bile that is stored in the gallbladder, detoxify the blood, synthesize protein, and store *glycogen* (blood sugar) and vitamins. The bile that is

gallbladder

stored in the _____ is necessary for the digestion of fat.

5 The gallbladder secretes into the duodenum, where its

bile

secretion, _____, emulsifies the fat in the chyme.

141

protein

glycogen; vitamins

liver

intestines

portal

hepatic

duodenum

retroperitoneal

6 The liver also synthesizes _____, stores blood sugar, or _____, and stores _____ as part of the digestive process.

7 The detoxification of the blood of substances such as alcohol takes place in the _____.

8 Blood reaches the liver for detoxification via the portal system of veins. These veins coming from the _____ empty into the sinusoids (veinlike compartments) of the liver.

9 The detoxified venous blood and other venous blood is collected into the hepatic veins, which empty into the major vein of the abdomen, the vena cava. Thus there is a _____ vein coming into the liver from the intestines and a _____ vein leaving the liver for the vena cava.

10 The pancreas is an exocrine and an endocrine gland with its head tucked into the horseshoe of the first part of the small intestine, the _____, and a tail that extends to the left to touch the spleen.

11 The pancreas lies behind the peritoneum and is therefore _____.

12 The duct of the pancreas releases enzymes into the duodenum that digest protein, fat, and carbohydrate. The fat has

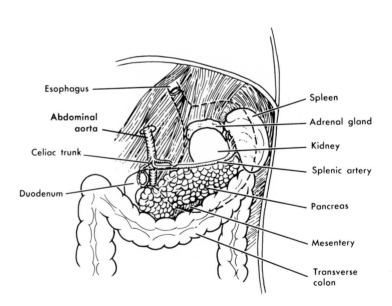

FIG. 11-2. Anterior view, stomach removed, showing the spleen, kidney, suprarenal (adrenal) gland, and pancreas. Note the hiatus in the diaphragm to admit the esophagus and one for the abdominal aorta.

bile

previously been emulsified by the _____ from the gallbladder.

13 The other foods digested by pancreatic enzymes are

protein; carbohydrates

_____ and _____ .

14 The endocrine function of the pancreas is to release a hormone known as *insulin,* which helps the body cells gain

glycogen

their blood sugar, or _____ , from the blood.

15 When there are insufficient amounts of the hormone

insulin

_____ released by the pancreas, a condition known as *diabetes* results.

16 The spleen is connected not with the digestive system but

circulatory

with the _____ system and lies on the left side against the diaphragm above the left kidney.

17 The spleen (Fig. 11-2) is somewhat variable in size and shape, weighs about 150 grams, and is actually a lymphatic organ somewhat like a lymph node. Like such an organ, its job is to

filter

act as a filter. The spleen acts as a lymph node to _____ out degenerating blood cells.

18 As a lymphatic organ, the spleen not only filters but also stores blood and produces antibodies from its position

left

superior to the (right/left) kidney.

12 The kidneys and suprarenal glands

1 The kidneys and ureters along with the bladder are part of the urogenital system responsible for maintaining the proper chemical composition of the blood. The kidneys of the

urogenital
_____ system lie on the posterior abdominal wall on the psoas muscles.

2 The ureters are collecting tubes that conduct the urine to the

ureters
bladder. The _____ , or collecting tubes, run down and forward to the bladder, which is situated in the anterior part of the pelvis.

3 The adrenal glands are also called *suprarenal glands* because

superior
they lie on the (superior/anterior/inferior) surface of the kidneys.

4 The kidneys lie so that the upper half of each is protected by the last two ribs, putting the left kidney in contact with the spleen and stomach and the right in contact with the largest

liver
organ in the body, the _____ .

5 The kidneys (Fig. 12-1) are about 12 cm. long and consist of an outer cortex (= bark) and an inner medulla (= marrow).

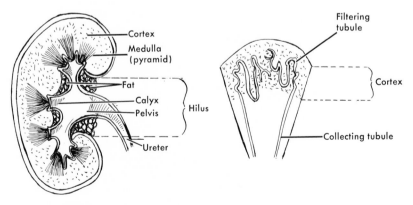

FIG. 12-1. *Left,* sagittal section of a kidney. *Right,* an enlargement of the cortex area to show the filtering tubules.

medulla	The inner portion, or _____ , contains the functioning unit of the kidney, the *nephron*.
	6 There are about a million nephrons in each kidney, and they are arranged in pyramids whose base is adjacent to the outer
cortex	layer, or _____ , and whose apices meet at a *calyx* (plural = calyces = cups).
	7 The cortex under the microscope is revealed to hold many convoluted tubules that process the fluid portion of the blood, whereas the medulla contains the collecting tubes,
calyces	which meet at one of several cuplike _____ .
urine	**8** The fluid waste, termed _____ , is gathered at the pelvis of the kidney, where the calyces meet, and is directed into the ureter.
	9 The ureter leaves the kidney at the hilus and travels about 10 inches down to the bladder. From the point where it leaves
hilus	the kidney at the _____ , the ureter narrows and has a thick muscular wall and a narrow lumen.
thick	**10** The ureter has a (thin/thick) muscular wall in order to pass
bladder	the urine within it down to the _____ by waves of contraction.
	11 The suprarenal, or adrenal, gland is actually two endocrine glands. Like the kidney, each gland has an outer layer, or
cortex; medulla	_____ , and an inner layer, or _____ .

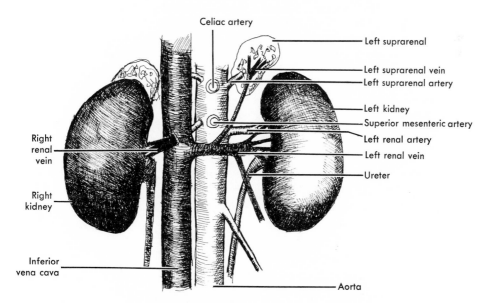

FIG. 12-2. Blood supply to the kidneys and suprarenals.

adrenal

12 The suprarenal, or _____ , gland secretes a steroid type hormone from its cortex and epinephrine from its medulla.

steroid

13 The hormone from the cortex, a _____ type hormone, aids in the metabolism of protein and maintenance of the electrolytic balance in the body.

epinephrine

14 From the medulla, the hormone called _____ acts to supplement the effects of the sympathetic nervous system, that is, prepare the body for "flight or fight."

15 The blood supply to the area of the kidney and suprarenal gland (Fig. 12-2) is via the *renal* artery. The venous drainage is

renal

via the accompanying vein, the _____ vein.

16 Nerves to the kidney come from the *celiac plexus* (parasympathetic) and *splanchnic* nerves (sympathetic). Sensory nerves accompany the splanchnic nerves. The suprarenals, however, since their secretions are essentially part of the autonomic

parasympathetic

nervous system, do not receive any (parasympathetic/sympathetic) nerve fibers because epinephrine acts in the same fashion as these fibers, that is, prepares the body for physical activity.

celiac

17 The parasympathetic fibers from the _____ plexus also go to the ureters, as do the sympathetic fibers from the

splanchnic

_____ nerves.

13 The respiratory system

The parts of the respiratory system are the nose, pharynx, larynx, trachea, bronchi, and lungs.

superior

1 Examine Fig. 13-1 and note that the pharynx is (superior/inferior) to the larynx.

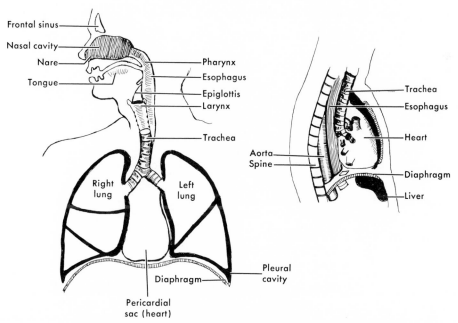

FIG. 13-1. *Left,* diagrammatic representation showing air and food passages and the lungs. *Right,* sagittal section showing the position of the trachea, esophagus, and heart.

greater

2 The right lung has a (greater/fewer) number of lobes than the left.

lungs

3 The trachea joins the pharynx to the _____ .

trachea

4 The epiglottis is part of the (esophagus/trachea).

anterior

5 The trachea is (anterior/posterior/inferior) to the esophagus.

6 The openings into the nose are called *nares*, or nostrils, and they are separated by the nasal *septum*. The wall separating

nares

the _____ is composed of three portions.

147

septum	**7** The first, or most anterior, portion of the _____ , or wall, is the *septal* cartilage; the second is a portion of the *ethmoid* bone; and the third is the *vomer* bone.
vomer	**8** The deepest segment of the septum is the _____
ethmoid	bone, and the middle is the _____ bone. The roof of the nasal cavity is also bone, the *cribriform* plate of the ethmoid.
	9 The bone in the roof of the nasal cavity called the
cribriform plate	_____ is pierced by several small foramina to give it a sievelike appearance.
foramina	**10** Through the several _____ in the cribriform plate pass the *olfactory* nerves, which contribute the sense of smell.
	11 From the lateral side of each nasal cavity project four shelves of bone called the *turbinate* bones, or *conchae* (Fig. 13-2).
lateral	These shelves projecting from the _____ side of each cavity increase the surface area over which the inspired air flows.

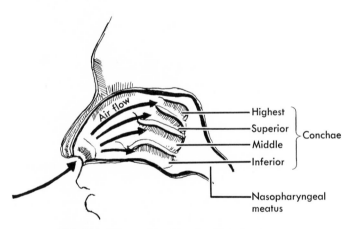

FIG. 13-2. The conchae through which the air passes.

concha (turbinate bone)	**12** Immediately below each shelf of bone, or _____
	_____ is a *meatus* named for the shelf above it.
below	Thus the inferior meatus lies (above/below) the
inferior	_____ concha.
	13 The meatuses open into the *nasopharyngeal* meatus, which, of course, leads to the windpipe, or trachea. The common meatus,
nasopharyngeal	or _____ meatus, thus receives the
four	combined airflow from (one/two/three/four) lesser meatuses.
	14 Surrounding the nasal cavity are various bones of the skull, some of which have *sinuses* (= hollow) that are important.

148

sinuses

The hollows, or _____ , are named for the bones in which they are located:

> maxillary sinus
> frontal sinus
> sphenoidal sinus
> ethmoidal sinus

maxillary; frontal

sphenoidal; ethmoidal

15 The four sinuses, the _____ , _____ , _____ , and _____ , are lined with a mucous membrane, and each has an opening that drains into the nasal cavity.

16 The purpose of the sinuses that drain into the

nasal cavity

_____ is obscure.

17 The conchae and walls of the nasal cavity are covered with mucous membrane richly supplied with blood. This

mucous

_____ membrane helps moisten the air passing over it while the rich blood supply warms the air.

18 The air leaves the common meatus called the

nasopharyngeal

_____ meatus and enters the *pharynx*, a tube common to both respiratory and digestive systems.

19 Thus the mouth can be used for breathing when the nose is plugged or when great amounts of air are needed (while running). (See Fig. 13-3.)

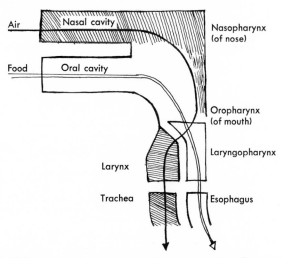

FIG. 13-3. Diagram illustrating how the trachea and esophagus coexist.

20 The larynx is a special apparatus sitting in the top of the

trachea

_____ , or windpipe, which has three functions.

larynx

21 The first function of the _____ in its position at the top of the trachea is to keep the air passages open.

keeping the air passages open

22 Another function of the larynx, in addition to its job of

_____ , is to prevent food or liquid from entering the trachea.

23 The third job of the larynx, in addition to keeping the air

preventing food from entering the trachea

passages open and _____

_____ , is to produce the sounds in speech and singing.

24 The larynx is part of the *oropharynx* and is composed of several pieces of cartilage. The larynx as part of the

oropharynx

_____ , makes up what is commonly called the *Adam's apple*.

25 Palpate your own Adam's apple (it is more prominent in males) and then swallow. Note that the Adam's apple moves

up

(up/down).

26 The *epiglottis* (Fig. 13-4) is pushed over the entrance of the larynx when one swallows in much the same manner as a lid closing on a box. The upward movement of the Adam's apple during swallowing is to tilt the epiglottis down.

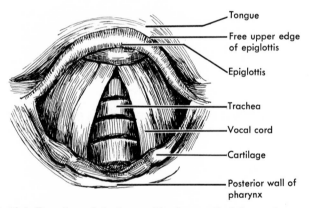

FIG. 13-4. Top view of the larynx. The front wall of the trachea can be seen between the vocal cords.

epiglottis

27 The _____ acts to close the larynx during swallowing, and the vocal cords at other times act to produce sound.

28 The vocal cords are operated by small muscles situated behind the palpable cartilage in the throat. These small muscles

vocal

operating the _____ cords are innervated by the *recurrent laryngeal* and *superior laryngeal* nerves. Both nerves are branches of the *vagus* nerve.

150

29 When one speaks, therefore, the muscles causing the vocal cords to vibrate properly are innervated by branches of the

vagus; recurrent laryngeal; superior laryngeal

_____ nerve, the _____

_____ and _____

nerves.

30 The trachea (= rough) itself is a tube about 15 cm. long with a diameter equal to that of a forefinger. The roughness of this

5

_____–inch long tube is due to rings of cartilage.

cartilage

31 The rings of _____ in the trachea are designed to reinforce it and maintain it as an open tube.

32 There are twenty cartilage rings designed to keep the trachea (open/closed), and they are actually horseshoe shaped, since they are incomplete on the posterior aspect.

open

33 The trachea divides into right and left *bronchi* in the chest.

bronchi

The right and left _____ lead into the lungs.

34 The right bronchus is larger than the left because the right lung is larger than the left. The cartilage reinforcement of the trachea is still evident in the early parts of the bronchi and, because of its size, more evident in the (right/left) bronchus.

right

35 The left lung is smaller than the right because the heart must be accommodated to the left side of the chest cavity.

36 The bronchus in the lung divides into *bronchioles* that ultimately become alveoli (= small hollows) (Fig. 13-5). The air in the pharynx proceeds through the larynx down the

trachea; bronchi

_____ , into the _____ , thence to

bronchioles; alveoli

the _____ , and finally to the _____ .

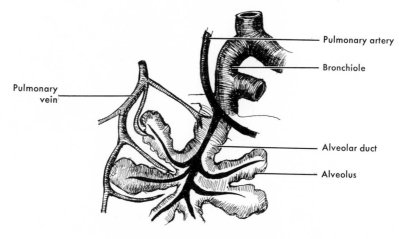

FIG. 13-5. The alveoli have a rich blood supply.

37 It is through the thin membranous walls of the

alveoli

_____ , or small hollows, that the gases are exchanged, that is, carbon dioxide for oxygen (Fig. 13-6).

38 Thus oxygen gets into the bloodstream by passing through

membranous

the thin _____ walls of the alveoli.

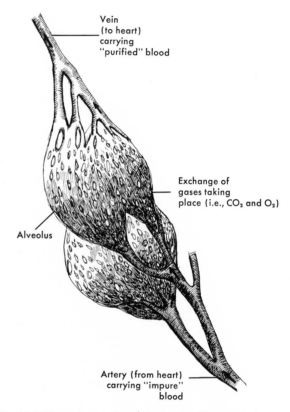

Vein
(to heart)
carrying
"purified" blood

Exchange of
gases taking
place (i.e., CO_2 and O_2)

Alveolus

Artery (from heart)
carrying "impure"
blood

FIG. 13-6. The exchange of gases in the alveoli is a gradual process.

39 The lungs (Fig. 13-7) can be considered to be somewhat like a huge cone-shaped sponge. Each lung becomes a half cone that

bronchus

has its air delivered via a right or left _____ at its center, or *hilum.*

hilum

40 At the center, or _____ , not only does the bronchus enter the lung, but blood vessels as well.

41 The top, or apex, of the lung is the thin end of the cone, and

apex

the inferior surface is domed. The top, or _____ of the lung fits behind the clavicle.

dome

42 The inferior surface of the lung is _____ shaped to fit the diaphragm.

152

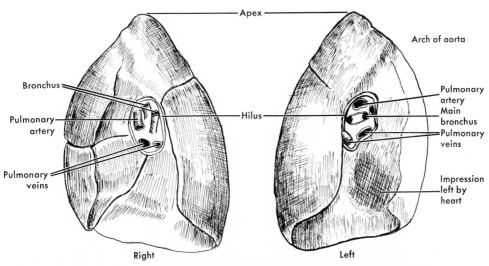

FIG. 13-7. Medial view of the right and left lungs. Note the position of the pulmonary arteries in relation to the pulmonary veins.

43 The lungs are divided into lobes, and each lobe is divided by septa of tissue. To reach the parts of each lung, the bronchioles must branch until the whole resembles a tree.

bronchioles

This treelike structure of the _____ , ensures that all areas of the lung are capable of getting air (although not all of them do in a sedentary person).

alveoli

44 To effect the exchange of gases in the _____ , the lungs are richly supplied with blood by the *pulmonary* (= lung) blood vessels.

pulmonary

45 The _____ blood vessels of the lungs are responsible for transporting "impure" blood to the lungs (the job of the arteries) from the heart and "purified" blood from the lungs (the job of the veins) to the heart.

46 The pulmonary artery thus carries "impure" blood to the

lungs

_____ , whereas the pulmonary vein carries "purified" blood to the heart. This is a reversal of the normal roles of arteries and veins.

153

$\mathbb{14}$ Circulation

THE HEART

1 The thoracic cage contains structures other than the lungs. The *mediastinum* (= standing in the middle), or septum, of the thoracic cage accommodates most of these other struc-

mediastinum

tures. Within the _____ , or septum, are the heart, blood vessels, trachea, esophagus, and *thymus* gland.

2 Among the structures in the mediastinum (Fig. 14-1) are the

heart

trachea, esophagus, thymus gland, _____ , and blood vessels.

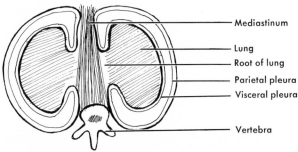

FIG. 14-1. The mediastinum. A horizontal section of the thorax shows the mediastinum and the two pleural cavities.

3 The heart is surrounded by a loose, tough outer coat called the *fibrous pericardium* (= around the heart). There is a

pericardium

similar inner coat called the *serous* _____ .

serous

4 The combination of an inner _____ peri-

fibrous

cardium and an outer _____ pericardium separated by a thin film of fluid permits the heart to beat in a nearly frictionless environment.

5 The heart is about the size and shape of the clenched fist of its owner and lies to the left of and behind the sternum. Palpate your own chest wall until a beat from the heart is

left

evident. Note that it is detected on your (left/right) side.

6 When making reference to the heart, think of it as your own, so that the right side, although it appears on the left part of Fig. 14-2, is truly the right side.

7 Examine Fig. 14-2 and note that the heart has two *atria* (= antechambers) and two *ventricles* (= little bellies). The flow of blood through the chambers is from (ventricle to atrium/atrium to ventricle/atrium to atrium).

atrium to ventricle

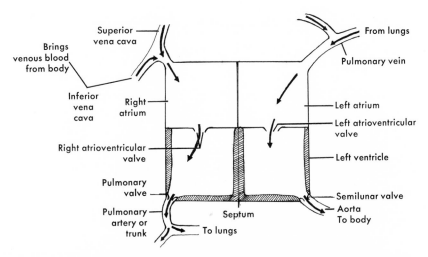

FIG. 14-2. Diagrammatic representation of the blood flow through the heart. Note that the blood vessels are in different positions (in reference to the individual chambers) from those of a real heart.

8 The right atrioventricular valve separates the right

atrium; ventricle   _____ and the right _____ .

9 The right atrioventricular valve is known as the *tricuspid* valve (cusp = point), and the left atrioventricular valve is known as the *bicuspid,* or *mitral,* valve. Thus the tricuspid valve is on

right   the (right/left) side of the heart, and the mitral valve is on the
left   (left/right) side.

left ventricle   10 The aortic valve separates the (left atrium and ventricle/left
  and aorta   ventricle and aorta).

body   11 The aorta takes blood to the (lungs/body/pulmonary vein).

venous (impure)   12 The vena cava contains (venous [impure]/arterial [pure])
     blood.

13 The blood vessels that transport blood from the heart to the

pulmonary arteries   lungs are the _____ .

14 The blood vessels that transport blood from the lungs to the

pulmonary veins   heart are the _____ .

15 The blood flow through the chambers of the heart and

right atrium; right ven-   through the lungs is as follows: _____ ,
  tricle; left atrium; left
  ventricle   _____ , the lungs, _____

     _____ , and _____ .

155

16 The thickness of the walls of the chambers varies. The left ventricle has the thickest muscular wall because it must pump blood to all parts of the body. Therefore the chamber of the

left ventricle
heart that does the greatest amount of work is the (right ventricle/right atrium/left ventricle).

thicker
17 The walls of the left ventricle are (thicker/thinner) than those of the other chambers (Fig. 14-3).

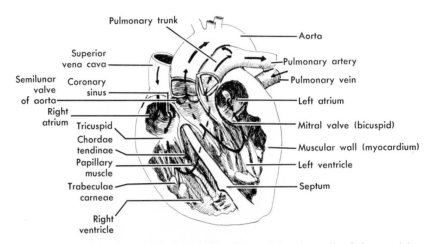

FIG. 14-3. Coronal section of the heart. Note the rough interior walls of the ventricles.

18 Before birth, the blood flows from right atrium to left atrium through the *foramen ovale*. The shock of birth closes this

foramen ovale
opening between the atria known as the _____

_____ , and the opening is marked by a small depression called the *fossa ovalis*.

fossa ovalis
19 The depression known as the _____ that marks the spot where the forament ovale was before birth is close to the *coronary sinus,* which returns venous blood from the heart muscle itself to the right atrium.

20 Thus there is venous blood returning to the right atrium from

vena cava
the body via the _____ and from the heart

coronary sinus
muscle via the _____ .

21 Attached to both atria are auricles (= ear). These appendages

auricle
have no great significance, but the term _____ , or ear, was formerly applied to the atrium itself.

22 When venous blood leaves the right atrium, it must proceed to

right ventricle; tricuspid
the _____ via the _____ valve.

23 The blood leaves the right ventricle via the *pulmonary valve* and enters the *pulmonary* (= lung) artery. Thus there are two

156

tricuspid

pulmonary

valves associated with the right ventricle, the _____

and the _____ .

24 The tricuspid valve is controlled by the *papillary* muscles, which are in close association with the *trabeculae carneae* (= beams of flesh), or ridges of muscle fiber in the walls. The

papillary

_____ muscles controlling the tricuspid valve are attached to the cusps by *chordae tendineae*.

chordae tendineae

25 The (trabeculae carneae/chordae tendineae) attach the papillary muscle to the cusps of the tricuspid valve. The papillary muscle prevents the cusps of the valve from everting or turning inside out from the pressure of blood within the ventricle.

26 The pulmonary semilunar valve closes by back pressure. The "pockets" of the cusps of the valve fill with blood, and this pushes the edges of the cusps together. No muscle structure is necessary.

27 Examine Fig. 14-4 and note that the pulmonary valve has (one/two/three) cusps that are half-moon, or semilunar, in shape. It is sometimes called the *pulmonary semilunar* valve.

three

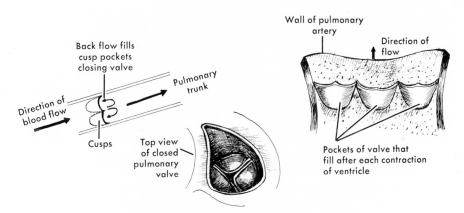

FIG. 14-4. Pulmonary valve.

semilunar

back pressure

28 The pulmonary _____ valve is caused to

close by _____ of the blood in the pulmonary artery.

29 The right and left pulmonary arteries lead directly to the

lungs

_____ .

30 In the lungs the blood loses some of its carbon dioxide and regains some oxygen in the capillary bed of the alveoli (Fig.

pulmonary
right

13-6). It then returns to the heart via the _____ veins to enter the (left/right) atrium. Examine Fig. 14-5 and note the relationships of the pulmonary vessels.

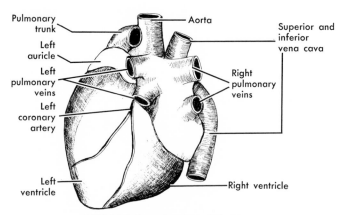

FIG. 14-5. Posterior view of the heart.

31 The passage of blood from left atrium to left ventricle occurs in the same way as on the right side of the heart. The valve between the left atrium and left ventricle is the

mitral; bicuspid _____ , or _____ , valve.

32 When the blood leaves the left ventricle through the aortic

semilunar _____ valve, it is under its greatest pressure because of the contraction of the powerful muscular walls of the left ventricle.

33 The operation of the aortic semilunar valve is identical to that of the pulmonary semilunar valve; thus it closes because of

back pressure _____ of the arterial blood.

34 The walls of the aorta inside the cusps of the semilunar valves have two small openings. When the "pockets" of the cusps are filled with blood, the openings receive a supply of blood. The

"pockets" openings within the _____ of the cusps of the aortic semilunar valve are the *left and right coronary* (= crown) arteries.

THE HEART MUSCLE

35 The two arteries that begin inside the pockets of the cusps,

left and right coronary the _____ arteries, supply the heart muscle itself (Fig. 14-6). Blood can enter these arteries only when the left ventricle is relaxed and the "pockets" are full.

36 The coronary arteries anastomose liberally to give the heart muscle a rich blood supply.

37 The aorta leaves the ventricle and arches above the heart. The arch of the aorta gives off major branches to the head and upper limbs. The arch of the aorta passes over the right pulmonary artery (Fig. 14-3).

158

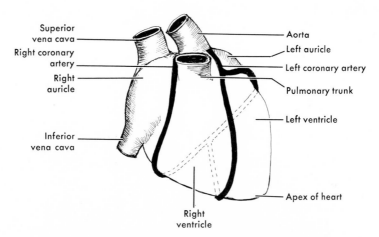

Superior vena cava
Right coronary artery
Right auricle
Inferior vena cava
Right ventricle
Aorta
Left auricle
Left coronary artery
Pulmonary trunk
Left ventricle
Apex of heart

FIG. 14-6. Right and left coronary arteries and their anastomoses.

38 The first blood vessels leading out of the aorta are the coronary arteries, and the second series of vessels are those to

head; upper limbs   the _____ and _____ .

39 The aorta then descends alongside the vertebral column as the descending aorta and passes through a hiatus in the diaphragm. At about this level it is behind the *apex* of the heart. Examination of Fig. 14-5 will reveal that the apex of the

left   heart is at the termination of the (left/right) ventricle.

40 The apex represents the end of the long axis of the heart running from behind, forward, and down to the left. The beat of the apex can be felt in the fifth interspace (between the fifth and sixth ribs) about 6 cm. to the left of the median plane. When one feels the apex beat, he can visualize the heart

obliquely   lying (horizontally/vertically/obliquely) backward from that point.

41 With the heart lying in this oblique position, the blood of

horizontally   necessity flows almost (horizontally/vertically) from the atria to the ventricles.

42 The average adult resting heart rate is about seventy beats per minute. This beat, or contraction, of the heart muscle is controlled by the *sinuatrial node,* a knot of tissue controlled in part by the vagus (brake) and sympathetic (accelerator)

seventy   nerves, which help maintain the _____ beats per minute.

sinuatrial   **43** The _____ node, which is linked to the

vagus; sympathetic   _____ and _____ nerves, is situated in the right atrium near the entrance of the superior vena cava.

44 The beat is spread from the sinuatrial node to the *atrioventricular* node and from there via *Purkinje* fibers to other

159

parts of the heart. Thus when the vagus nerve stimulates the
heart to (speed/slow), the stimulation travels from the

slow

atrioventricular

node to the _____ node and finally to the

Purkinje

_____ fibers.

speed

45 The sympathetic nerves cause the heart to (speed/slow), and
the contraction of the ventricles, termed *systole*, is (in-
creased/decreased).

increased

systole

46 When the _____ , or contraction of the ventricles,
has ceased, the atrioventricular valves open, and blood is
pumped from the atria to the ventricles. This pumping action
is called *diastole* and refills the ventricles.

47 The heart muscle has two contractions. The first, called
systole, pushes blood into the aorta and pulmonary arteries,

diastole

and the second, called _____, pushes blood
into the ventricles.

in sequence

48 The atria and ventricles thus contract (simultaneously/in
sequence/at any time).

49 A contraction of a ventricle ejects about 60 ml. of blood into

seventy

the body's general circulation. At _____

60

beats per minute, this _____ ml. of blood ejected at
each contraction totals about 4 liters per minute. A well-
conditioned athlete's heart may pump about twice this

8

amount, or _____ liters per minute.

MAJOR ARTERIES

50 Examine Fig. 14-7 and note that the first major artery to run

brachiocephalic

from the aorta is the innominate, or _____ ,
artery.

51 The term *brachiocephalic* means that the artery goes both to

arm; head

the brachium, or _____ , and the _____ . (Trace
the other artery branching from the innominate.)

52 The second artery running from the aorta is the left common

carotid

_____ artery.

53 The third artery branching from the aorta is the left

subclavian

_____ artery.

54 The abdominal aorta divides into right and left common

iliac

_____ arteries.

55 The common iliac artery becomes the major artery of the

femoral

lower limb, the _____ artery.

56 The companion vein to the common cartoid artery is the

internal jugular

_____ vein.

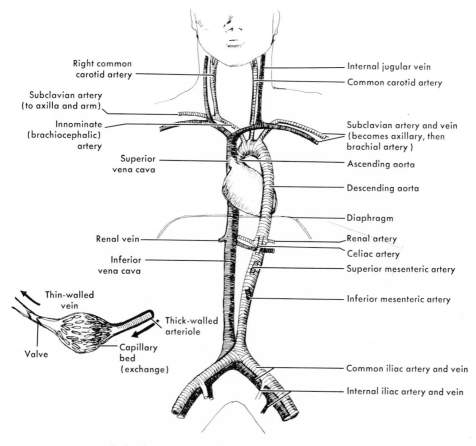

FIG. 14-7. Major arteries of the body and a capillary bed.

57 There are four major arteries shown in Fig. 14-7 as supplying blood to the viscera of the abdomen. They are the celiac, the

renal; superior mesenteric; inferior mesenteric

_____ , the _____ , and the _____ arteries.

58 The walls of all arteries are thicker and more muscular than

more

those of veins because the arteries must withstand (more/less) pressure.

MAJOR VEINS

59 Since the veins are less muscular and blood within them has only about 5% of the original pressure with which it left the heart, another mechanism is necessary to control the blood

low

flow. The flow of blood in the veins under (high/low) pressure is aided by one-way valves not unlike the pulmonary and aortic valves.

60 Examine the veins in your own arm in the area of the elbow by applying pressure above the elbow. Note that the large lateral vein, the *cephalic,* and another medial vein, the *basilic,* are joined in some variable fashion across the elbow joint by the *median cubital* vein. The latter is used for taking blood samples, giving transfusions, etc.

61 Most veins are paired with arteries and have the same nomenclature. Among the exceptions are the three veins of

cephalic
basilic; median cubital

the upper limb, the _____ ,

_____ , and _____ ,

and the *azygos* (= unpaired) system (Fig. 14-8).

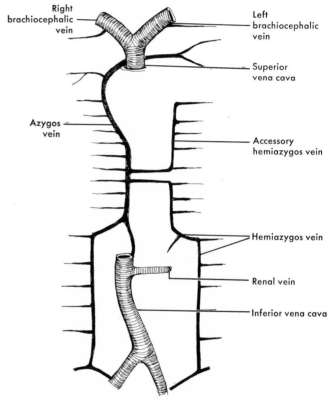

FIG. 14-8. Veins of the thorax and abdomen. Note that these veins are not part of the vena cava system until the azygos vein empties into the superior vena cava.

62 The azygos system is an alternate route to the vena cava for
right
blood returning to the (right/left) atrium.

azygos

63 The _____ , or unpaired, system has two other

hemiazygos
venous branches, a lower one called the _____

accessory azygos
vein and an upper one called the _____

_____vein.

LYMPHATICS

64 The lymphatic (= clear water) system (Fig. 14-9) is an

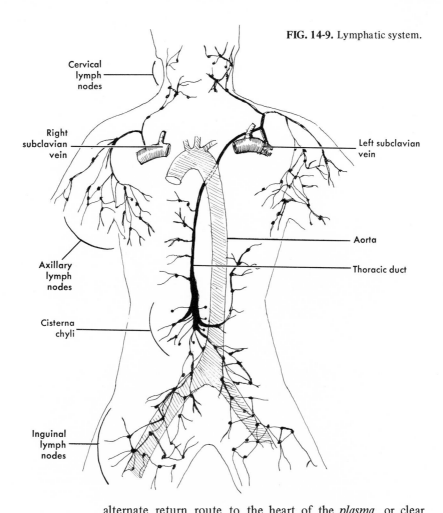

FIG. 14-9. Lymphatic system.

Cervical lymph nodes

Right subclavian vein

Left subclavian vein

Aorta

Thoracic duct

Axillary lymph nodes

Cisterna chyli

Inguinal lymph nodes

alternate return route to the heart of the *plasma,* or clear portion of the blood. The lymph vessels carrying the

plasma   _____ portion of the blood are very delicate and thin walled.

65 The flow of plasma or lymph is very slow but is increased in activity. The flow is filtered by dense masses of tissue called *lymph nodes,* which are concentrated in three areas: the cervical, axillary, and groin regions.

nodes   66 The filters, or _____ , situated in the

cervical; axillary   _____ , _____ , and
groin

_____ regions filter out substances such as bacteria and also produce lymphocytes.

67 The largest lymph vessel, the *thoracic duct,* empties into the left subclavian vein. The thoracic duct begins at the *cisterna chyli,* then proceeds up the thorax to empty into the

left subclavian   _____ vein.

15 The pelvis and contents

1 The so-called true pelvic cavity (Fig. 15-1) is the funnel-shaped space within the pelvis that contains the lower parts of the

ureters

alimentary canal, the bladder, the _____ leading into the bladder, and parts of the genital, or sex, organs.

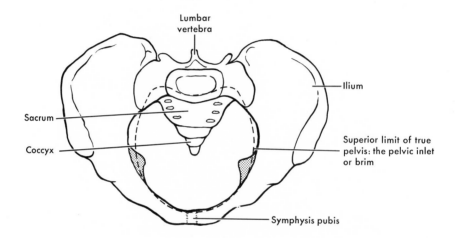

FIG. 15-1. Top view of a male pelvis. The pelvic inlet is marked; the outlet lies directly inferior to it.

2 The *lumbosacral* joint unites the vertebral column with the sacrum at the fifth lumbar vertebra. The sacrum is also joined

coccyx

at its inferior end to another bone formation, the _____.

3 Several strong and important ligaments running from the

lumbosacral

ilium to the vertebral column support the (lumbosacral/sacrococcygeal) joint.

4 The sacrococcygeal joint is also reinforced by strong

ligaments

_____.

5 The joint between the sacrum and the ilium known as the *sacroiliac joint* is a synovial joint that is stabilized by both ligaments and bony irregularities. The ligaments are named

sacroiliac

as the joint is. They are the _____ ligaments.

symphysis pubis | **6** The joint at the front of the pelvis, the _____

_____ , is a cartilaginous joint and is very stable, although the joint is more moveable in women during childbirth.

7 The walls of the pelvic cavity are covered by the obturator internus muscle laterally and by the piriformis and coccygeus posteriorly. Both the muscle on the lateral walls, the

obturator internus | _____ , and the

piriformis | muscles on the posterior, the _____

coccygeus | and the _____ , have a great many important circulatory and nerve trunks crossing them.

8 The floor of the pelvic cavity is composed of fascia and muscle tissue that form first a pelvic and then a urogenital diaphragm. The pelvic diaphragm is superior to the

urogenital | _____ diaphragm (Fig. 15-2).

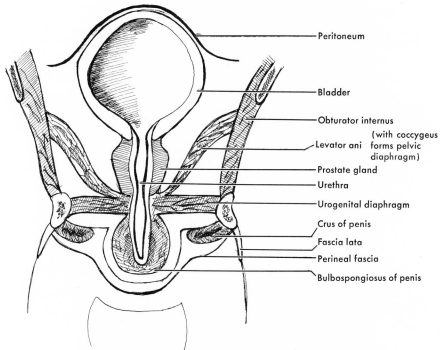

FIG. 15-2. Frontal section of a male pelvis and urogenital region.

Labels: Peritoneum, Bladder, Obturator internus (with coccygeus forms pelvic diaphragm), Levator ani, Prostate gland, Urethra, Urogenital diaphragm, Crus of penis, Fascia lata, Perineal fascia, Bulbospongiosus of penis

pelvic | **9** Both the _____ diaphragm and the

urogenital | _____ diaphragm forming the floor of the pelvic cavity support the organs in the cavity such as the

alimentary | lower parts of the _____ canal, the

165

bladder

urinary

urethra

pubic

muscle

trigone

mucosa

trigone

four

20

prostate gland

_____ of the urinary system, and the genital organs.

10 The bladder of the _____ system lies on the pubis within the pelvis.

11 The urethra is the passage from the bladder to the exterior. Thus the bladder has three openings: two ureters leading into it and the _____ leading out of it.

12 The bladder itself is a hollow, thick-walled organ that sits in the pelvic cavity against the _____ bones.

13 The thick walls consist of interlacing bundles of involuntary muscle. The bladder is capable of great stretching by virtue of the _____ bundles that make up its walls.

14 The interior mucosa of the bladder is wrinkled and folded except in an area called the *trigone*. The trigone is a triangular area whose angles are formed by the urethral orifice and the orifices of the ureters. In this triangular area known as the _____ the mucosa is smooth.

15 The trigone is immobile because it is fastened down to the urogenital diaphragm. Thus the bladder inflates like a balloon as it fills with urine (at least 200 ml. is necessary to elicit the desire to urinate), and the folds in the interior _____ disappear, whereas the smooth _____ remains unchanged.

16 The male urethra is about 20 cm. long, extending from the bladder through the prostate gland, the pelvic diaphragm, the urogenital diaphragm, and the penis. The urethra passes through (three/four/two) areas on the course of its _____ cm. length.

17 The female urethra is about 4 cm. long and passes through the pelvic and urogenital diaphragms and opens in front of the vagina. The female urethra does not pass through the (pelvic diaphragm/prostate gland/urogenital diaphragm).

166

16 The perineum and the reproductive systems

THE PERINEUM

1 To understand the relationship of the genital organs a diamond-shaped area below the pelvic diaphragm known as the *perineum* must be examined. The diamond-shaped

perineum

_____ is divided into two triangles.

2 The two triangles of the perineum (Fig. 16-1) are formed by drawing a line joining the two ischial tuberosities. There is thus an anterior, or urogenital, triangle and a posterior, or

urogenital

anal

anal, triangle. The anterior, or _____ , triangle differs in construction from the posterior _____ triangle.

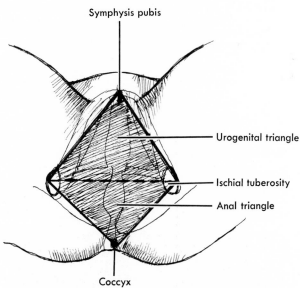

Symphysis pubis

Urogenital triangle

Ischial tuberosity

Anal triangle

Coccyx

FIG. 16-1. The perineum, the urogenital and anal triangles.

3 The urogenital triangle of the male contains the external

167

urogenital	genitalia and the urogenital diaphragm. The _____ diaphragm in the urogenital triangle is composed of superficial (weak and poorly developed) and deep transverse perineal muscles.
	4 The muscles of the urogenital diaphragm, the superficial and
transverse perineal	deep _____ muscles, run from the ramus of each ischium to the perineal body.
	5 The perineal body is the tendinous center of the diamond-shaped perineal area. It is an area of muscle attachment for the superficial and deep transverse perineal muscles that arise
ischial	on each _____ bone.
	6 The urogenital triangle of the female contains the external genitalia, the lower end of the *vagina,* and the urethra, as well
urogenital	as the _____ diaphragm.
	7 The two muscles of the urogenital diaphragm, the superficial
transverse perineal	and deep, _____, muscles are fused into a single muscle in the female. The female diaphragm is divided into two halves by the two orifices it contains, the
vagina	urethra and _____ .
	8 The anal triangle is covered by a perineal membrane. The triangle is bounded anteriorly by the posterior edge of the urogenital diaphragm and the tendinous center, the
perineal	_____ body.
	9 The anal triangle is roofed over by only the pelvic diaphragm, or levator ani muscle, and has thus a fossa on each side of the rectum known as the *ischiorectal fossa.* The roof of each fossa
levator ani	is formed by the (gluteus maximus/levator ani/transverse perineal) muscle.
anal	**10** The ischiorectal fossa (Fig. 16-2) in the (urogenital/anal) triangle is filled with fat that gives support to the rectum but can be displaced during defecation.

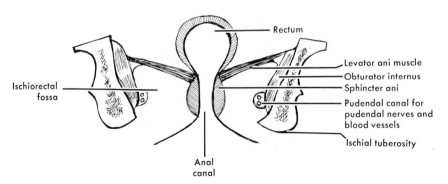

FIG. 16-2. Coronal section of the pelvis showing the ischiorectal fossa.

MALE REPRODUCTIVE SYSTEM

11 The male genital organs (Fig. 16-3) consist of paired testes and a penis and the tubes known as the *vas deferens* connecting them. Along the course of the tubes and the

testes

urethra are several glands. The paired _____ and the penis are external, and the conducting tubes, the

vas deferens

_____ , are largely inside the pelvis.

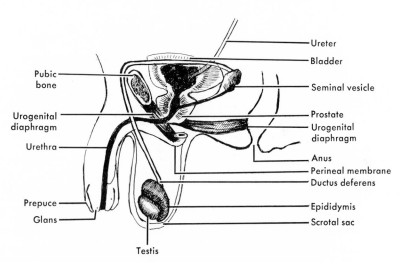

Pubic bone
Urogenital diaphragm
Urethra
Prepuce
Glans
Testis

Ureter
Bladder
Seminal vesicle
Prostate
Urogenital diaphragm
Anus
Perineal membrane
Ductus deferens
Epididymis
Scrotal sac

FIG. 16-3. Scheme of the male genitalia.

12 Each testis originates in the abdominal cavity and descends by the eighth month of life to the scrotum, or sac, dragging with it over the pubic bone representatives of each layer it passes

scrotum

through. Each testis lying in the _____ thus has a long covered cord leading up into the body cavity.

13 The cord leading up into the body cavity is the *spermatic* cord, which contains a duct, nerves, and blood vessels. The

spermatic

_____ cord pierces the abdomen just above the pubic bone at a weakening in the abdominal wall called the *superficial inguinal* ring, often the site of hernia (Fig. 10-5).

14 The spermatic cord, containing blood vessels, nerves, and a

duct

_____ , is responsible for the transmission of sperm from the testis through the abdominal wall at the

superficial inguinal ring

opening called the _____ to the urethra.

15 The duct for conduction of sperm is termed the *ductus deferens,* or *vas deferens.* Note on Fig. 16-3 that the

<table>
<tr><td>ductus deferens</td><td>sperm-conducting tube, the _____ , runs across the bladder and forms the seminal vesicle.</td></tr>
</table>

16 The ductus deferens has a sac, or diverticulum, in it at the level of the bladder called the *seminal vesicle*. This diverticulum,

seminal vesicle

the _____ , secretes a fluid into the ductus deferens to facilitate the transport of the sperm.

17 The prostate gland lying below the bladder receives the ductus deferens as it joins the urethra there. The

prostate

_____ gland, like the seminal vesicle, secretes a fluid that is added to the sperm.

inferior

18 The urethra pierces the prostate, which lies (superior/inferior) to the bladder, and then passes through the urogenital diaphragm.

19 Inferior to the urogenital diaphragm is the *perineal* membrane, which the urethra pierces to reach the penis. Because

perineal

this section of the urethra pierces the _____ membrane, it is termed the membranous urethra.

20 The urogenital diaphragm controls the flow of urine in the

sphincter

urethra by acting as a guard muscle, or _____ , on the urethra.

21 The section of the urethra in the penis is called the *penile,* or *spongy, urethra.* This section of the urethra traverses the

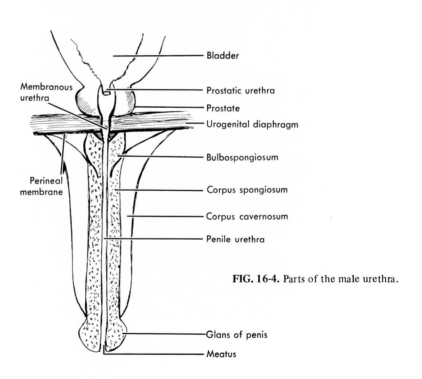

Membranous urethra

Perineal membrane

Bladder

Prostatic urethra

Prostate

Urogenital diaphragm

Bulbospongiosum

Corpus spongiosum

Corpus cavernosum

Penile urethra

Glans of penis

Meatus

FIG. 16-4. Parts of the male urethra.

170

length of the *corpus spongiosum,* hence its name—the penile,

spongy

or _____ , urethra.

22 The corpus spongiosum (Fig. 16-4) is the central of three cylindrical erectile tissues in the penis (= tail). The paired cyl-

corpus spongiosum

inders on either side of the _____ are the *corpora cavernosa* (singular = corpus cavernosum).

23 The corpora cavernosa are anchored to the inferior pubic rami near the ischial tuberosity and then pass forward and

corpora cavernosa

converge. The paired cylinders, the _____

_____ , lie above the central cylinder con-

corpus spongiosum

taining the penile urethra, the _____

_____ .

inferior pubic rami

24 The corpora cavernosa are attached to the _____

_____ and at this point of attachment are called the *crura* of the penis.

25 The corpus spongiosum is attached to the perineal membrane by the bulb of the penis. The corpus spongiosum proceeds

perineal

from the bulb attached to the _____
membrane to the glans (= acorn) at the end of the penis.

26 The corpus spongiosum is covered by muscle where it is

bulb

attached at its proximal end at the _____ of the penis. The muscle is called the *bulbospongiosus.*

27 The corpora cavernosa are also covered by muscle at their proximal attachments. The muscle surrounding each corpus cavernosum is the *ischiocavernosus,* found at the attachment

inferior pubic rami

of the crura on the (perineal membrane/inferior pubic rami).

28 The spongiosum and the cavernosa dilate with blood to maintain an erection during intercourse. The muscle struc-

ischiocavernosus

tures at their proximal ends, the _____
muscle on each corpus cavernosum and the

bulbospongiosus

_____ muscle on the corpus spongio-
sum, act to prevent the venous return from the three cylinders, which are engorged with blood during intercourse.

29 The penis is covered with skin, and at the distal end, the

glans

acorn-shaped _____ is partly covered by a moveable cuff of skin known as the *prepuce* (= foreskin).

FEMALE REPRODUCTIVE SYSTEM

30 The paired ovaries of the female are analogous to the paired

testes

_____ of the male reproductive system. They arise near the kidneys and migrate downward into the pelvis.

171

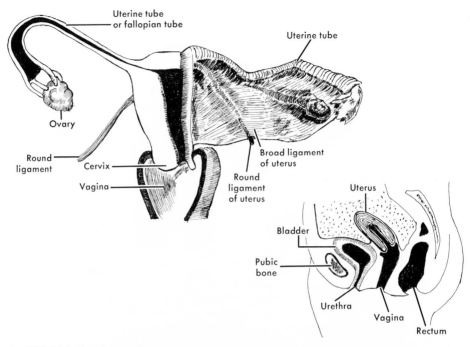

FIG. 16-5. Female reproductive system. Front view partially dissected and a median section.

pelvis

31 The ovaries (Fig. 16-5) situated in the _____ release a single egg once per month. This egg drops into the uterine tube, which runs medially to the uterus.

32 The uterus is a thick-walled, pear-shaped organ whose stem points down between the bladder and the rectum. The uterus

laterally

thus receives two uterine tubes that run (anteriorly/laterally/medially) from it.

bladder

33 The uterus, situated between the _____ in

rectum

front and the _____ behind, leads into the vagina.

34 The uterus has a round ligament attached to it on each side that passes outside the pelvic cavity by going through the abdominal wall. In proceeds downward and merges with the external genitalia. The round ligament of the uterus is thus

spermatic cord

analogous to a structure on the male, the _____

_____ .

35 The uterine tube is held by a double fold of peritoneum called the *broad ligament* that merges on the uterus. The broad ligament thus is pierced by another ligament, the

round

_____ ligament of the uterus.

36 The vagina commences at the *cervix,* or entrance from the uterus, and terminates at the external genitalia. A glance at

backward
forward

Fig. 16-5 will show that the vagina is tilted (forward/backward), whereas the uterus is tilted (forward/backward).

37 The lower end of the vagina is covered by the labia (= lips)

cervix

minora. The distal end of the uterus, the _____ , is encompassed by the vagina, whereas the distal end of the

labia minora

vagina is encompassed by the _____ .

38 The labia minora (Fig. 16-6) run forward to meet at a small lump of tissue known as the clitoris, the center of sexual excitement in the female. The labia minora are situated in the

urogenital

(anal/urogenital) triangle.

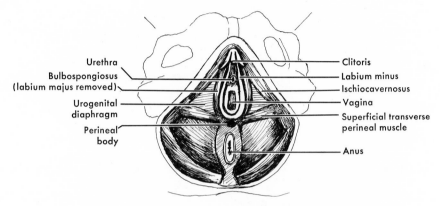

Urethra
Bulbospongiosus
(labium majus removed)
Urogenital diaphragm
Perineal body

Clitoris
Labium minus
Ischiocavernosus
Vagina
Superficial transverse perineal muscle
Anus

FIG. 16-6. Detail of the female genitalia.

clitoris

39 The center of sexual excitement, the _____ , and the labia minora are both covered by the labia majora.

40 The external urethral orifice is found just posterior to the

labia minora

clitoris, between the (labia minora/labia majora).

41 The muscles of the external genitalia of the female are analogous to those of the male. There will thus be paired

ischiocavernosus

_____ muscles originating near the ischial tuberosities and running forward and a

bulbospongiosus

central _____ muscle.

42 The paired ischiocavernosus muscles run to the clitoris. The

penis

clitoris is therefore analogous to the _____ in the male.

43 The bulbospongiosus muscle in the female proceeds forward

perineal

from its attachment at the central _____ body and surrounds the vaginal opening.

44 The blood supply to the perineal area (Fig. 16-7) comes from branches of the internal iliac artery, chiefly the internal pudenal (= shameful parts) and the external pudenal, a branch of the femoral artery. The internal pudendal artery

within

begins (within/without) the pelvic cavity.

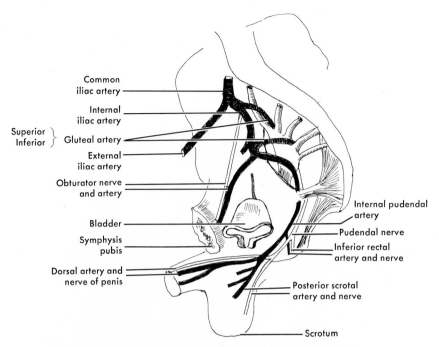

FIG. 16-7. Blood supply and innervation of the genitalia and perineum.

pelvic

pudendal

parasympathetic

45 The innervation of the perineal area must provide sensory, motor, sympathetic, and parasympathetic effects. Thus the rectum is supplied by (pelvic/abdominal) splanchnic nerves; the external genitalia are supplied by a nerve bearing the same name as the major artery of the area, the _____ artery; and the remaining parts have sympathetic fibers supplied from plexuses in the area.

46 The splanchnic nerve provides (parasympathetic/sympathetic) fibers. The pudendal nerve, arising from S2, S3, and S4, supplies most of the innervation to the perineum. Its fibers are primarily motor and sensory.

17 The brain and nervous system

THE BRAIN

1 The nervous system can be divided into two parts: the *central* and *peripheral* systems. The central nervous system is composed of the brain and spinal cord, and the second

peripheral

division, the _____ system, is made up of the many spinal nerves and twelve cranial (= of the skull) nerves.

twelve

2 The spinal nerves and the (ten/twelve/eleven) cranial nerves of the peripheral system also make up the autonomic system, which controls the so-called automatic functions such as breathing, digestion, and heart rate.

automatic

3 The autonomic nervous system, controlling the _____ functions of the body, is really a subdivision of the peripheral nervous system.

central

4 The brain (Fig. 17-1) of the (central/peripheral) nervous system has three major divisions: the *cerebrum,* the *cerebellum* and the brainstem.

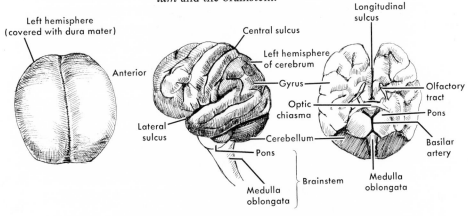

Left hemisphere (covered with dura mater)

Anterior

Central sulcus

Left hemisphere of cerebrum

Gyrus

Optic chiasma

Lateral sulcus

Cerebellum

Pons

Medulla oblongata

Brainstem

Longitudinal sulcus

Olfactory tract

Pons

Basilar artery

Medulla oblongata

FIG. 17-1. The brain. Top, side, and end views.

cerebrum

5 The cerebrum, or forebrain, has two hemispheres deeply wrinkled like a walnut and is the largest part of the brain. The folds of the forebrain, or _____ , serve

175

to increase the surface area. They are termed *gyri* (=bent).

6 The cerebrum has a cortex, or outer coat, and an inner layer,

cortex   or medulla. The outer layer, the _____ , is

gyri   made up of gray matter with many folds, or _____ .

7 The gray matter is composed of the bodies of nerve cells whose *axons* (= axle) pass into the inner portion, the

medulla   _____ , which is prinicipally made up of the white matter of the long threadlike *axons*.

8 Within the medulla there are gray centers of matter inter-

axons   spersed among the white tendrils, called _____ , which come from the nerve cells of the cortex.

medulla   9 The fibers within the _____ of the cere-brum run from one fold to the next or to more distant folds or down to the brainstem. The valleys between the folds are called *sulci* (singular = sulcus).

10 The cerebral hemispheres, especially the cortex, are the centers of highest mental and behavioral activity; they also control muscular activity and contain sensory mechanisms.

cerebrum   Thus the hemispheres of the _____ are responsible for most of our conscious functions.

below   11 The cerebellum (= little brain) is situated (above/below) the cerebrum and is the center of coordination of movement. It thus functions closely with the cerebrum when the latter is

muscular   concerned in (mental/muscular) activity.

12 In performing a front roll, a gymnast initiates the contraction

cerebrum   of the many muscles concerned in the _____ of the brain. The movements caused by these contractions

cerebellum   are coordinated in the _____ .

13 Certain area of the cerebral cortex control specific muscular functions. About a hundred such areas are known.

anterior   14 The motor areas are (anterior/posterior) to the central sulcus (Fig. 17-2).

15 The brainstem is principally composed of the *pons* (= bridge) and the *medulla oblongata*. The main function of two parts of

pons; medulla ob-   the brainstem, the _____ and the _____ longata

_____ , is to control such involuntary func-tions as respiration and circulation.

CRANIAL NERVES

16 The cranial nerves connect the brainstem to those parts of the body serving the involuntary functions such as

respiration; circulation   _____ and _____ .

176

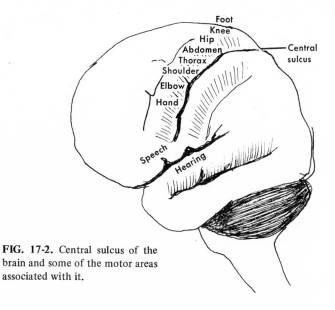

FIG. 17-2. Central sulcus of the brain and some of the motor areas associated with it.

Table of cranial nerves

NUMBER	NERVE	CHIEF FUNCTION
1	Olfactory	Sense of smell
2	Optic	Sense of vision
3	Oculomotor	Movement of eye; constriction of eye pupil
4	Trochlear	Movement of eye
5	Trigeminal	Sensation of head and face; movement of jaw
6	Abducent	Movement of eye
7	Facial	Taste; movement of face; secretion of tears
8	Vestibulocochlear	Hearing; balance
9	Glossopharyngeal	Taste; movement and sensation of pharynx; secretion of saliva; visceral reflexes
10	Vagus	Taste; sensation in throat and bronchial tree; movement in throat and larynx; movement of digestive tract; visceral reflexes
11	Accessory	Movement of throat, head, and shoulders
12	Hypoglossal	Movement of tongue

17 There are twelve pairs of cranial nerves containing various amounts of different fibers. Some fibers are sensory only, whereas others are sensory and motor. The twelve

cranial

_____ nerves supply all the special organs of such functions as sight, hearing, and taste, as well as supply the viscera. (See Table of cranial nerves.)

18 The vagus nerve, the tenth cranial nerve, for example, is responsible for much of the control of the digestive tract. The innervation supplied by the vagus is (sympathetic/parasympathetic).

parasympathetic

19 The mnemonic for the cranial nerves is "on old Olympus' towering top a Finn viewed Germans vaulting a hop."

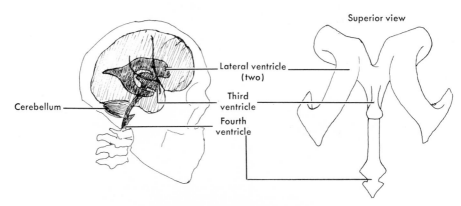

FIG. 17-3. Ventricles of the brain.

VENTRICLES AND CEREBROSPINAL FLUID

20 Within the brain itself are cavities known as *ventricles* (Fig. 17-3), which are filled with fluid. There are four of these

ventricles

cavities, or _____ , in the brain—all interconnected.

four

21 The ventricles are filled with cerebrospinal fluid. All (four/three/two) ventricles secrete this fluid from specialized capillaries known as *choroid plexuses.*

choroid plexus

22 The cerebrospinal fluid produced by the _____

_____ of the ventricles flows very slowly through the ventricles until it reaches an area below the cerebellum. Here it seeps out of the fourth ventricle.

fourth
below

23 The cerebrospinal fluid seeps out of the (first/fourth/second) ventricle in an area (above/below) the cerebellum. At this point it oozes upward around the brain under one of the brain's coverings and downward around the spinal cord under an extension of the same brain covering, the *arachnoid* (= spider web) *membrane.*

arachnoid

24 The cerebrospinal fluid under the _____ membrane is absorbed by the veins and venous sinuses of the brain and the veins of the spinal cord.

25 The absorption of the cerebrospinal fluid by the

venous

_____ system of the brain and spinal cord is very slow. The probable function of the fluid is to act as a

178

liquid cushion to protect the delicate nervous tissue.

26 The function of the cerebrospinal fluid produced in the

ventricles; protect

_____ is to _____ the
delicate nervous tissue of the brain and spinal cord.

27 The roof over each lateral ventricle (one in each cerebral
hemisphere) is called the *corpus callosum* (callosum = hard)

white

and is composed of (white/gray) fibers that maintain com-
munication between the hemispheres.

corpus callosum

28 The roof of the lateral ventricles is the _____

_____ , and the base of the cerebral hemispheres
is made up of masses of gray matter termed the *corpus
striatum.*

29 The fibers that leave the cerebral hemispheres for the
brainstem pass through the base of the hemispheres, the

corpus striatum

_____ .

THE THALAMUS AND HYPOTHALAMUS

30 The thalamus (Fig. 17-4) (plural = thalami) is also found at the
base of the cerebral hemispheres. It is a sensory center second
only to the cerebral cortex. This secondary sensory area, the

thalamus

_____ , is responsible for relaying to
the cerebrum sensations produced by movements of the
muscles, tendons, and joints.

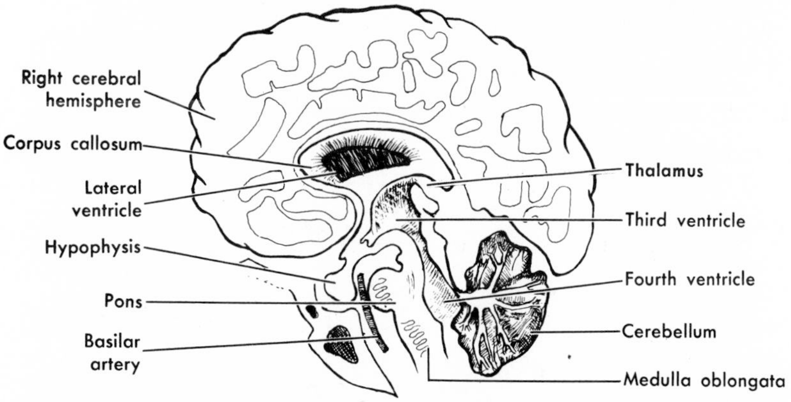

FIG. 17-4. Sagittal section of the brain showing the interconnections of the ventricles.

31 The thalamus is responsible for relaying sensations from the

muscles; tendons;
joints

_____ , _____ , and _____ ;
the general area below it, the *hypothalamus,* is an autonomic
center regulating smooth muscle, cardiac muscle, and glands.

32 The area below the thalamus (Figs. 17-1 and 17-4), the

hypothalamus

_____, includes the *optic chiasma* and
hypophysis.

179

33 The optic chiasma, as its name implies, is concerned with the sense of (sight/hearing/smell), and in this area a nerve from each eye crosses (chiasma = X).

34 The second important component of the hypothalamus is the

_____, sometimes called the *pituitary gland.*

35 The hypophysis, or _____ gland, is an endocrine gland important for secreting growth hormone, thyroid-stimulating hormone, luteinizing hormone, and others.

36 The *pons* is part of the (cerebellum/cerebrum/brainstem) and gets its name from its bridgelike structure and function.

37 The fibers of the pons cross at right angles to the direction of the brainstem (Fig. 17-5, *A*). Where these right-angle fibers of

the _____ enter the cerebellum, they are known as *cerebellar peduncles* (= feet).

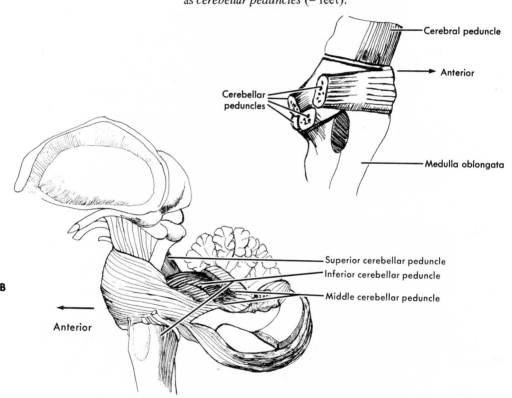

FIG. 17-5. **A,** Pons and brainstem. **B,** Cerebral peduncles.

38 There are also peduncles carrying bundles of fibers from the cerebral hemispheres through the pons down to the spinal

cord. They are known as _____ *peduncles* (Fig. 17-5, *B*).

39 Most of the fourth ventricle lies in the pons except for the tip, which lies within the structure immediately inferior to

medulla oblongata

the pons, the _____ .

40 The medulla oblongata tapers to become the spinal cord. The medulla oblongata carries the major efferent (going to the muscles) pathways of the central nervous system from the

cerebrum

motor center above it in the _____ to its

spinal cord

extension below, the _____ .

41 In the floor of the fourth ventricle lying in the

medulla oblongata

_____ are centers for regulation of respiration, body temperature, and heart rate.

MENINGES

42 The brain and spinal cord are surrounded by three membranes, or meninges (Fig. 17-6): the *dura mater,* the *arachnoid,* and the *pia mater.* Of the (two/three/one) meninges, the

three

dura mater is the outermost.

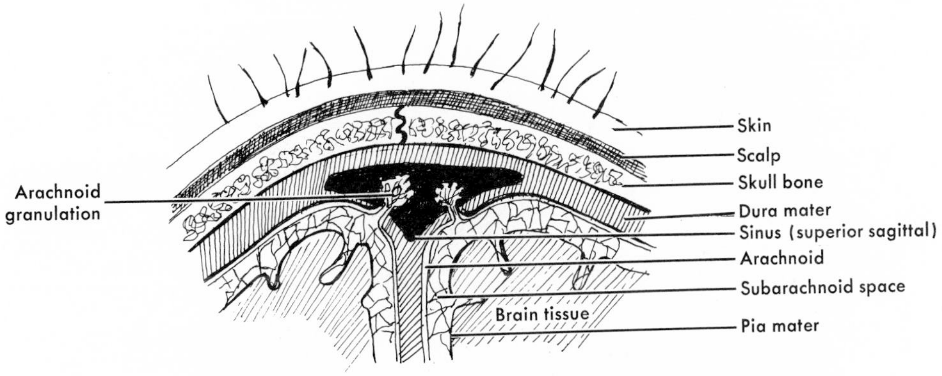

FIG. 17-6. Coronal section of the meninges, brain. The meninges continue onto the spinal cord.

meninge

43 The dura (= hard) mater, the outermost _____ , or membrane, is thick and tough and supports the soft brain tissue. It also forms partitions that divide the cranial cavity into compartments.

44 In addition to its job of dividing the cranial cavity and lending

support

_____ to the soft brain tissue, the dura mater contains large venous sinuses that receive the venous drainage of the brain.

45 The arachnoid membrane, the middle of the three membranes, is very thin and delicate. The chief function of the delicate

arachnoid

_____ membrane is to absorb cerebrospinal fluid.

46 Structures called *arachnoid granulations* that project into the dural sinuses are believed to be responsible for the absorption

cerebrospinal of the _____ fluid.

pia **47** The innermost of the meninges is the _____ mater.

48 The pia mater covers the brain very closely, following the

gyrus contours of each raised fold, or _____ , and

sulcus going down into each crease, or _____ . The
other two meninges do not do this.

49 The function of the pia mater is to form the choroid plexus in

cerebrospinal each ventricle for the production of _____
fluid.

50 The brain tissue itself is insensitive. The pain of headaches
comes from the blood vessels or from the outer meninge, the

dura mater _____ .

BLOOD SUPPLY TO THE BRAIN

51 The blood supply to the brain (Fig. 17-7) comes from two *internal carotid* arteries and two *vertebral* arteries. The fact that

four there are (four/two/three) major sources of blood ensures an
adequate supply to the brain.

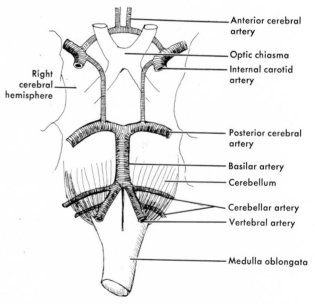

FIG. 17-7. Arterial circle of the brain.

52 The four anastomose at the base of the brain. The anastomo-

internal carotid sis of the most anterior of these arteries, the _____

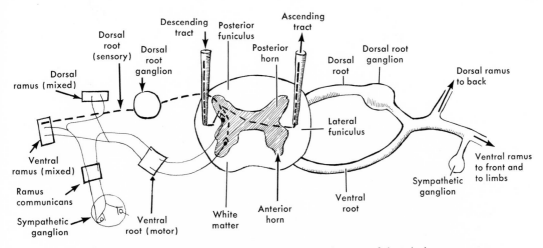

FIG. 17-8. Horizontal section of the spinal cord showing the roots of the spinal nerves.

_____ arteries, with the basilar artery forms an arterial circle (the circle of Willis) at the base of the cerebral hemispheres (Fig. 17-7).

SPINAL CORD

53 The medulla oblongata becomes the spinal cord (Fig. 17-8) as it emerges from the *foramen magnum,* a large opening in the skull. The spinal cord is covered by the same three meninges

dura mater   that cover the brain, the _____ ,
arachnoid; pia mater

_____ , and _____ .

54 The spinal cord stretches from the opening in the skull, the

foramen magnum   _____ , down a variable distance. It stops anywhere from T12 to L3. The remainder of the vertebral canal is filled with a long leash of nerves called the *cauda equina* (= horse's tail).

55 The nerves at the inferior end of the spinal cord making up

cauda equina   the "horse's tail," or _____ , inner-
lower   vate the (upper/lower) limb.

56 There are thirty-one pairs of spinal nerves, and each one emerges from the bony spinal column through an intervertebral foramina. Each of the (twenty-nine/thirty/thirty-one)
thirty-one   pairs of nerves is covered by dura mater and arachnoid, both of which gradually fade away as the spinal nerve is prolonged.

57 The spinal nerves issue from the spinal column through

intervertebral foramina   _____ and are numbered according to their vertebrae. The cervical nerves issue above their corresponding vertebrae until the eighth cervical verte-

183

brae. From C8 to L5 the spinal nerves issue below their corresponding vertebrae.

58 The third lumbar spinal nerve leaves the spinal column (above/below) the third lumbar vertebra.

below

59 The eighth cervical nerve leaves the spinal column (above/below) the eighth cervical vertebra.

below

60 The spinal cord from which the nerves originate is fashioned in such a manner that its structure is opposite to the general structure of the brain. That is, the (white/gray) matter of the spinal cord is on the inside, and the (white/gray) matter is on the outside.

gray
white

61 The gray matter is shaped like the letter H (Fig. 17-8), and each leg of the H sends two roots to join laterally to form a

spinal

_____ nerve.

62 The gray matter contains tracts of fibers from the brain. The motor control area of the brain in the _____ hemispheres sends motor or efferent fibers down.

cerebral

63 The motor, or _____ , fibers travel downward, whereas the sensory or afferent fibers travel upward.

efferent

64 The descending _____ , or _____ , fibers and the ascending _____ , or _____ , fibers are also in company with fibers from the autonomic nervous system.

motor; efferent

sensory; afferent

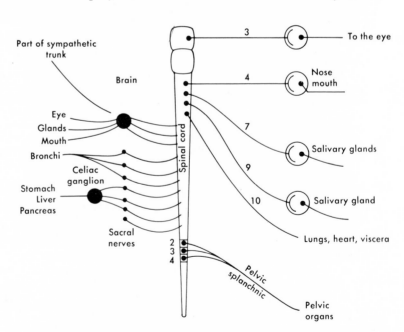

FIG. 17-9. Scheme of the craniosacral outflow (the parasympathetic nervous system) and the thoracolumbar outflow (sympathetic nervous system).

65 The autonomic nervous system (Fig. 17-9) has two portions: the sympathetic, known as the *thoracolumbar* outflow, and the

parasympathetic

_____ , known as the *craniosacral* outflow.

thoracolumbar

66 The sympathetic nervous system, or the _____ , outflow, forms a series of ganglia on the inside body wall just lateral to the vertebral column. These paired chains of ganglia are known as the sympathetic trunk.

trunk

67 The ganglia forming the sympathetic _____ stretch from the first cervical vertebra to the coccyx.

craniosacral

68 The parasympathetic nervous system, or _____

_____ outflow, includes separate cranial nerves and some sacral nerves.

69 The smooth muscle of the digestive system, the heart muscle, and the glands of the body are controlled by the two divisions

autonomic

of the _____ nervous system.

70 The areas controlled by the autonomic nervous system such

digestive system; heart

as the _____, _____ ,

glands

and _____ send afferent, or sensory, impulses to the spinal cord, where they are transmitted to either the sympathetic or parasympathetic system.

71 The afferent (sensory) nerves trigger response in the efferent (motor) nerves of either the sympathetic or parasympathetic systems. The sympathetic system prepares the body for "flight or fight," whereas the other system, the

parasympathetic

_____ system, prepares the body for "rest and repose."

flight or fight

72 The preparation for "_____" by the sympathetic nervous system is enhanced by the release of *epinephrine* from the suprarenal, or adrenal, glands. This hormone prepares the body for vigorous physical activity.

73 When a person is startled by a loud noise and he jumps up

sympathetic

with heart beating, eyes wide, etc., his _____ nervous system has obviously been stimulated.

afferent
efferent

74 One can see that a spinal nerve contains sensory or (afferent/efferent), fibers, motor, or (afferent/efferent), fibers, as well as fibers from the autonomic nervous system.

75 The efferent impulses from the cortex of the

cerebrum

_____ can follow two pathways down the spinal cord. One is called the *pyramidal* system and the other the *extrapyramidal* system.

185

76 The pyramidal system is the more direct route, and the

extrapyramidal

second pathway, the _____ system, is indirect.

77 The pyramidal system commences in that part of the cerebral

anterior

hemispheres controlling motor function located just (anterior/posterior) to the central sulcus.

78 From the motor center the fibers travel down via the cerebral peduncle to the most inferior part of the brainstem, the

medulla oblongata

_____ .

79 From the medulla oblongata, the fibers descend in the white matter in the area known as the lateral *funiculus* (= rope). In the spinal cord the fibers proceed from the white matter to

gray

the _____ matter, where they synapse (Fig. 17-8) with the spinal roots in the anterior horn.

80 The one pathway, then, for motor, or efferent, impulses is

cerebrum

from the cortex of the _____ , to the

medulla oblongata

_____ , in the brainstem, to the lateral

funiculus; anterior

_____ of the spinal cord, to the (anterior/posterior) horn of the spinal gray matter.

81 The extrapyramidal pathway for motor impulses is less direct. From the cerebral cortex, the impulse goes to the area of the

pons

brainstem called the *bridge,* or _____ . From here, it is sent to the cerebellum via the cerebellar peduncles (Fig. 17-5).

82 The impulse travels to the cerebellum via the

cerebellar

_____ peduncles, and here it is coordinated with others.

83 The impulse now travels upward to the midbrain to an area

peduncle

called the *red nucleus* via yet another _____ .

red

84 From the _____ nucleus in the midbrain, the impulse descends into the white matter of the spinal cord in the same areas that the pyramidal fibers descend into, the

lateral funiculus

_____ .

85 Thus the extrapyramidal pathway begins in the cortex of the

cerebrum; pons

_____ proceeds to the bridge, or _____ ,

cerebellum

and is sidetracked to the coordinating center, the _____ ;

red

it then ascends to the _____ nucleus in the midbrain;

lateral funiculus

from here it descends into the _____ of the white matter.

86 Also in the brainstem are fibers that descend from a network

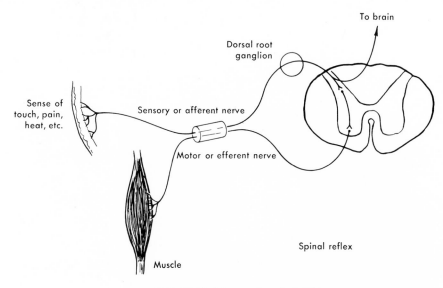

To brain

Dorsal root
ganglion

Sense of
touch, pain,
heat, etc.

Sensory or afferent nerve

Motor or efferent nerve

Spinal reflex

Muscle

FIG. 17-10. Scheme of a spinal reflex.

of cells called the *reticular system.* This system, too, is extrapyramidal and like the extrapyramidal pathway is

efferent

concerned with (afferent/efferent) impulses.

87 There are thus three ways by which motor impulses descend

pyramidal

from the brain, the _____ , the

extrapyramidal; reticular

_____ , and the _____ systems.

88 Another important aspect of motor function is movement in emergencies. This is controlled by the spinal reflexes (Fig.

afferent

17-10) and involves both types of fiber, the _____

efferent

and the _____ fibers.

89 The afferent, or sensory neurons have their cell bodies in an enlargement of the dorsal root called the *dorsal root*

ganglion

_____.

90 The impulse telling you that your hand is on a hot surface runs to the dorsal root ganglion, from where it may go to the

ascending

(ascending/descending) tract in the lateral funiculus or to the anterior horn.

anterior

91 By going directly to the _____ horn, the sensory impulse triggers a group of muscles that will move your hand very quickly.

92 The message will reach your brain via the ascending tract, but by that time you have moved your hand, since the anterior

motor

horn of the spinal gray matter is concerned with (sensory/motor) nerves that act to move your hand.

187